Boris Vladimirowich Silvestrov

Energy for the Ages Gravity

AF534340

Boris Vladimirowich Silvestrov

Energy for the Ages Gravity

Alternative energy

ScienciaScripts

Imprint

Any brand names and product names mentioned in this book are subject to trademark, brand or patent protection and are trademarks or registered trademarks of their respective holders. The use of brand names, product names, common names, trade names, product descriptions etc. even without a particular marking in this work is in no way to be construed to mean that such names may be regarded as unrestricted in respect of trademark and brand protection legislation and could thus be used by anyone.

Cover image: www.ingimage.com

This book is a translation from the original published under ISBN 978-620-7-46825-6.

Publisher:
Sciencia Scripts
is a trademark of
Dodo Books Indian Ocean Ltd. and OmniScriptum S.R.L publishing group

120 High Road, East Finchley, London, N2 9ED, United Kingdom
Str. Armeneasca 28/1, office 1, Chisinau MD-2012, Republic of Moldova, Europe
Printed at: see last page
ISBN: 978-620-7-74092-5

Copyright © Boris Vladimirowich Silvestrov
Copyright © 2024 Dodo Books Indian Ocean Ltd. and OmniScriptum S.R.L publishing group

Contents

Keywords: two-way pontoon pump, piston pump with
fixed piston, Pelton hydroelectric turbine, gravity, alternative energy.

About the author

Born on 05.01.1953. In 1976 he graduated from Azerbaijan Polytechnic Institute on speciality "Refrigeration and compressor equipment and air conditioning" He worked in power engineering. He was engaged in repair of various power equipment. I am familiar with energy problems according to my working life. Alternative energy at first as an interest, as a hobby. After making a demonstration model of a sea wave power plant and retirement, together with the model he got a job in the Azerbaijan Research Institute of Geotechnical Problems of Oil, Gas and Chemistry in the laboratory of alternative energy as an engineer. While working in the laboratory, he created four alternative energy technologies, which can be found in this book.

Introduction

My passion for alternative energy started in the early 2000s. At that time it was like a hobby. The huge flow of information on this topic aroused in me an interest in the problem of harnessing the energy of the sea wave. This energy was presented in mass media as depending on the length of the sea wave front. Calculations were given as to what the final value of this energy could be. But it was impossible to find even a hint of how it could be harnessed. By education I am an engineer mechanic on refrigeration and compressor installations, but it so happened that I worked in power engineering for a considerable part of my working life. I repaired generators and other power equipment at thermal and hydroelectric power plants. He also repaired generators on an offshore oil platform. Before retirement, I worked as a service mechanic for the last ten years in the Azerbaijan-Finnish company WARTSILA, on service maintenance of diesel modular power plants. My knowledge and my professional experience played a decisive role in the possibility of solving the problem of harnessing the power of sea waves. While still working at WARTSILA, during downtime, with the permission of my superiors, I started to make a demonstration model of a sea wave power plant as it was in my imagination. This work took me more than two years. By that time I had already, as it seemed to me, basically found a way to harness the energy of the sea wave. And it was not wave energy, but gravitational energy. The height of the wave and its periodic change only contributed to the work of the mechanism capable of receiving electricity from the gravitational field. And that mechanism was a two-way piston pump, a pontoon pump with a
with a fixed piston. It is the design of the two-way fixed piston pontoon pump that I have constructed that, when the water level is cycled, makes it possible to generate clean energy in a variety of ways that will be proposed in this book.
In my first paper, it is about an alternative energy plant that converts the energy of sea waves into electrical energy. It's basically a marine hydroelectric power station. It's my imagination. So far, there is nothing like it in the world, but it may well become a reality. In many ways, it is similar to a conventional hydroelectric power station. This idea was realised in the form of a demonstration model, which will be described below. For a more complete perception of this idea, a project of a pilot installation of a sea wave power plant was created. Having understood the fundamental approach to solving the problem of converting the energy of sea waves into electricity by the material of the prototype, it is easy to apply this approach to the design of a full-fledged sea wave power plant. As we know, the capacity of any hydroelectric power plant, is determined by two parameters. It is the water head, which is measured in

metres of water column, where 10m of water column is equal to the pressure of one atmosphere, and the water flow rate - measured in m^3 /sec. The design I have proposed fits this definition perfectly. It provides the hydroelectric units with a given, necessary amount of water and necessary head. But, unlike conventional hydroelectric power plants, where the water flow rate is a finite value, for each particular river, it has the undeniable advantage that for offshore hydroelectric power plants, the water flow rate can be any required value, determined by the number of working pumps in this wave power plant. And this is an undeniable advantage over conventional hydroelectric power plants, given that the construction cost per MW produced is much lower than the construction cost of conventional hydroelectric power plants. And besides the head can be selected in tens and hundreds of metres of water column. There are only a few such adequate high-rise dams in the world. And if we take into account that in combination with such a head, the water flow rate can be of any size, then the capacity can be planned of any size. And, taking into account that the construction cost per one megawatt of electricity produced by offshore hydroelectric power plants is much lower than that of conventional energy sources, offshore hydroelectric power plants will certainly occupy a dominant position in the world energy sector in the future. If the location of such a plant is chosen in such a way that the coastline is the boundary between the sea and a high, mountainous shore, it is advisable to design it together with a pumped storage plant. The pressure that can be created by a two-way pontoon pump is 250-300atm. And this is enough to raise the water to a considerable height. Part of the cylinders will directly pump water into the reservoir located in the mountains. Also, the excess water, after the compensation columns, will not be discharged into the sea, and too, will be pumped into the reservoir. As the wave amplitude decreases, the power will be compensated by the operation of the hydro-storage station. Taking into account meteorological observations for the previous years, and with correct calculations in choosing the number of working cylinders,

working at the plant will ensure a stable energy supply. This plant is designed to have eight working cylinders. But in principle they can be as many as you like, Let's stop on the chosen eight-cylinder variant, and the diameter of the piston equal to 2 m. The project calculates one slave cylinder, designs its device and describes its operation.

This offshore hydroelectric power plant is a platform installed on the shelf, similar to offshore oil production platforms. It is located on supports made in one piece with guide grids for the working cylinders. The working grids are necessary to ensure that the cylinders move in a strictly vertical direction. The

grids are metal structures made of channels and are capable of carrying significant lateral loads from sea currents and wave impacts. On all four sides of the inner part of each guide grating there are chutes in the form of two channels on which the support rollers welded to the sides of the pontoon sections run. In this way, the section can only move up and down with the crest and trough of the wave. And it does not matter what the direction of the waves is. Nor does it matter what the level of the water horizon is at low tide or high tide. This change in water level is taken into account when designing the dimensions of the station and the possibility of moving the movable cylinder relative to the stationary piston. The above-water part of the plant is made in the form of two tiers, although a single-tier version is possible, but with a larger area to accommodate all the planned equipment, working and living quarters. The "support legs" are thick-walled pipes of large diameter, made as a whole with all the guide grids, generally having a great weight and, consequently, great stability. The first tier is located at 8m above sea level. The pump room provides the hydroelectric units with the necessary amount of water. The second tier is a generator hall provided with an overhead crane of 200 tonnes lifting capacity. It houses the main power equipment necessary for the operation of a conventional hydroelectric power plant. Adjacent to it are utility and production facilities. The pumping compartment, as well as the generator compartment has an overhead crane with lifting capacity of 200t. The cranes are designed for maintenance and repair work. The presence of operating personnel and qualified repairmen allows these offshore power plants to be serviced in normal mode, just like conventional hydroelectric power plants. In relation to all currently proposed analogues of marine wave energy converters, this design has an undeniable advantage. Firstly, all electrical equipment is located above the water level, and secondly, and most importantly, the ability to promptly intervene in the production processes, including service and repair right on site, by repairing individual components and units, without stopping the entire plant.
Let's consider the scheme of operation of this installation. From the pontoon pumps, water under high pressure, which can reach several hundred atm, is supplied to the compensation columns. The pressure is selected in such a way that pipelines and all working equipment can withstand it. This is about 250 atm. Pipelines of superheated steam at thermal power plants work under such pressure, but in more severe hot conditions, so we will consider that such pressure can be chosen as working pressure in this case. This is the operating pressure of the expiring jet when it hits the blades of the bucket hydro unit. Such a high pressure can be obtained due to the fact that the pontoon pump is also a hydraulic press. The diameter of the working cylinder is many times greater than

the total diameter of the expelling jets hitting the blades of the hydraulic unit. In principle, the pressure obtained by the pontoon pump can exceed 250 atm by many times. But in the working scheme there are at least two pressure regulators. The first one is designed to provide a stable operating pressure for water supply to the blades of the hydraulic turbine. The second pressure regulator plays the role of a safety valve. It discharges water back into the reservoir when the operating pressure exceeds a pre-selected value. At the beginning, the water is pumped into the compensation column. This column is designed to partially equalise the pulsating nature of this pumping process. At the outlet of the column itself, on the common manifold, as already mentioned, there are pressure regulators. The compensation column is a vertical vessel operating under high pressure. The upper part of the column is filled with air and the rest with water. This air acts as a spring and equalises the pulsating nature of the pressure created by the piston pump.

As the wave height increases, the water flow rate increases and the pressure increases. As the wave height increases, the water flow rate increases and the pressure rises. This causes the next pressure regulator to come into operation, adjusted to a pressure slightly higher than that of the previous regulator. The excess water is discharged into the sea. There is virtually no ecological damage caused by this installation. The water intake openings of the pontoon pumps have grids to prevent marine life from entering the process network. In addition, if necessary, the entire installation can be fenced around the perimeter with fine mesh. Technogenic accidents with catastrophic consequences, such as hydroelectric dam failures, are completely excluded. And even if we imagine the possibility of destruction of an offshore hydroelectric power plant due to tsunami, sabotage, military actions, there will be practically no ecological damage.

The platform is equipped with a cargo berth and a conventional berth. The cargo berth is provided by an overhead crane. Pontoon sections and other equipment can be repaired either on site or delivered for replacement by sea transport. The domestic berth is designed for delivery and changeover of operating and maintenance personnel. There are two cargo lifts on the installation, designed to facilitate the movement of operating personnel and small loads across the platform. At the topmost point of the platform there is a helicopter pad, which is also used for delivery and removal of operating personnel from the platform. There are rescue floats and mechanisms for their delivery to the water surface. These mechanisms are special winches working both in automatic and manual mode. The roof of the platform can be used as a resting area for working personnel

The location of the platform is chosen so that the bottom relief is gentle and provides for the arrival of high waves. According to calculations, this installation is mounted at a depth of 20m. Countries that have ocean coasts where waves reach five metres or more can get electricity at very low prices. Below in the text there are photos and description of a representative model of a sea wave power plant.

Photo No. 1

1) The first picture shows the engine room of the pump room. On the left and right side there are four pump section support covers (yellow) each. On the left side there are three assembled covers and one dismantled cover.

The dismantled cover indicates that the first pump unit has been taken out for repair. The upper support cover of the pump section is on special stands after dismantling. There are two compensation columns in the far background. Water from the pumps enters the compensating columns, (blue vertical apparatuses) then, passing through the pressure regulator, also blue in colour on the pipeline line is directed to the generator, machine room. This room is located above the pump room. The shut-off valves allow any required unit to be repaired. In the pumping engine room there is an overhead crane with a lifting capacity of 200 tonnes. It is designed for equipment maintenance and repair work.

2) In the second photo (below the text), there is the second tier of the offshore wave power plant. In the middle are two hydroelectric units with a Pelton turbine, to which water is fed from below from the pump room. To the left of them are two transformers to prepare the transmission of electricity by sea cable to the mainland. As well as the pump room, there is an overhead crane (200 tonnes) for maintenance and repair work. There is a helipad on the upper roof of the generator room. There is a staircase to the pad from the common roof. The roof itself is a resting place for repair and maintenance personnel. In the right part of the second tier there are three-storey household compartments and a

control room. A lift shaft is visible at the rear. On the sides there are pedestrian passages and stairs for transition between floors. After exhausting through a special pipeline the water is discharged back into the sea. This pipeline is clearly visible in photos #3 and #4 It is located vertically along the lift shaft.

Photo #2.

Photo 3 shows the generator engine room, as well as Photo 2. In addition to photograph No. 2, the water supply pipework to the hydraulic turbines and the waste water discharge pipework can be seen, which is located next to the second goods lift shaft between the pump and generator rooms. Part of the outrigger cantilever tracks for the pump room overhead crane can be seen on the left side. These cantilever tracks are intended for unloading of moored cargo ships.

Photo #4

Photo #4 shows almost the entire offshore hydroelectric power plant from the end face. At the bottom, the support legs on which the entire structure sits on the shelf soil can be seen. Underneath the pump house are the guide grids. They are fastened together and form a single structure with the whole station. The pontoon pumping sections of double-way pumps with fixed pistons move in a strictly vertical direction along these grids. On the left side, a cantilevered outrigger of crane tracks can be seen for overhead crane access to the free zone for loading and unloading operations from the deck of moored vessels. From the end of the pump room you can see the mooring platform for end mooring of cargo ships. Mooring cranes can be seen to soften the mooring process. From this platform there is an entrance to the lift to the generator room. Lift shaft on the left side of the façade

of the plane.

Photo #5

Photo No. 5, the same as photo No. 1, is a view of the pumping section of the offshore wave hydroelectric power plant. But unlike in this photo, in the left corner there is a pumping section prepared for replacement. The upper support cover of the first pumping section has been dismantled and installed on special supports. The pump section to be replaced is suspended on special supports in the form of an I-beam. On the sides of the pump section, there are guide rollers. These rollers move along the guide grid to ensure a strictly vertical movement of the pump section.

Photo #6 shows the offshore wave power plant at its full height The support legs are 20m high. The seawater does not reach the lower edge of the pump room base 8m. The photo clearly shows the guide grids assembled as a single unit. Inside them, the pumping sections move with the changing level of the sea wave, thus pumping seawater under high pressure into the compensation columns. At the accommodation level are automated winches for launching the necessary lifeboats. The lifeboats are suspended on special brackets, which operate in automatic and manual mode. On the front side, you can see the pipeline outlet from the expansion columns. Water from the compensation columns flows through pressure regulators to the generator room to supply water to the Pelton hydro turbines. Three floors of utility rooms can be seen above. The control room for the offshore wave hydroelectric power plant is also located there. This offshore platform, as well as offshore oil production platforms, is fully suitable for full operation and accommodation of maintenance and repair personnel

In photograph No. 7, at the floor level of the pump room, there is a mooring berth for the reception of personnel arriving by sea. Mooring cranes are visible

along the berth. There are emergency ladders throughout the entire height, allowing access to all levels of the offshore power plant in the event of a lift system failure. There are also platforms that allow travelling on parallel levels of the residential part of the power station. In the upper part you can see the helipad and the stairs leading to it. Helicopter transport of personnel as an alternative to sea transport of personnel. Helicopter transport is intended for emergency communication with the shoreline.

Photo #8

Photo #8 Rear of the offshore wave hydroelectric power plant. The lift shaft is located in the centre of the structure. From the sea berth, having climbed the stairs to one level, one can get to the entrance to the lift, from where it is possible to get to any level of the working part of the personnel location by lift. The same lift also provides access to the upper observation deck and from it to the helipad. All vertical and horizontal movements around the personnel working area are duplicated by stairs and transition platforms in case of emergency.

The basis for converting gravitational energy into electricity is a pontoon piston pump with a fixed piston. A precondition for its operation is a cyclic change in the water level. In fact, this pump is the "heart" of the mechanism for converting gravitational energy into electricity. Let's look at one of the possible designs of this pump, consider how it works, and familiarise ourselves with calculations of the power it can generate. All of this can be found in the article below:

CHAPTER 1

DESIGN AND CALCULATION OF A PROTOTYPE OFFSHORE, WAVE HYDROELECTRIC POWER PLANT.

Azerbaijan Research Institute of Geotechnical Problems of Oil, Gas and Chemistry
Author: Engineer of the Alternative Energy Laboratory
Boris Vladimirovich Silvestrov boris_boris_silvestrov@mail.ru@mail.ru

Project assignment

1) Installation depth 5-6m.
2) Tie-in to existing trestles (piers).
3) Minimal, cost-effective value.
4) Minimum overall dimensions.
5) Use of standardised units and equipment.
6) Wave height from 0.3m.

Purpose of the project assignment.

Calculation and design of an experimental power plant with its subsequent manufacture and testing of the created sample, to confirm the calculation part and the correctness of the selected design, which will allow to proceed to the construction of a full-fledged, marine hydroelectric power plant of any given capacity. The need for powerful, economically favourable alternative energy sources today is enormous. The proposed variant fully meets the demands of today, both in terms of simplicity of the proposed design and low cost resources. An offshore power plant is a device that converts the energy of sea waves into electricity. This conversion is carried out by a powerful pump, which is also a hydro-pressor, providing the hydro-turbine with the necessary water flow and pressure. The construction of the designed power plant will be divided into two phases. At the first stage, the pumping unit will be manufactured and assembled according to scheme No.1. And only after the calculated part of the project is confirmed, which consists in close indicators to the calculated ones, both for the created pressure and water flow rate, the second stage will be carried out - the purchase of the hydraulic unit itself.

Scheme #1

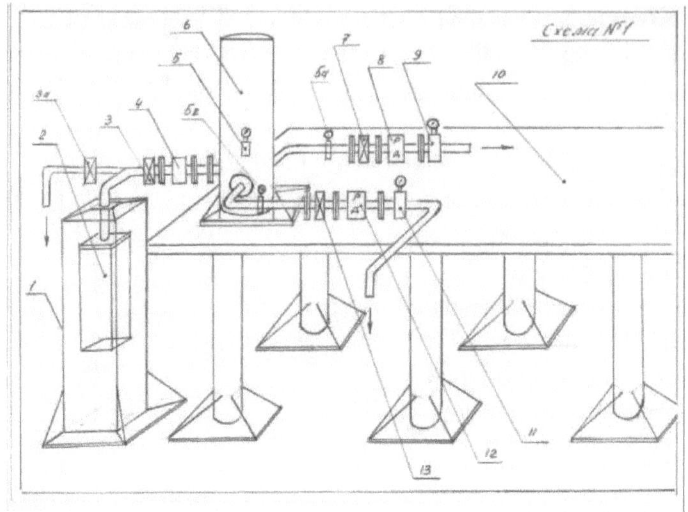

1) .Guide grid.
2) .Pontoon pumping section.
3) . Shut-off valves.

3.a). Shut-off valves on the process water discharge line

4) . Check valve.
5) . Pressure gauge.
5) a). Pressure gauge, on the water supply line to the hydroturbine blades
6) в). Pressure gauge, on the excess water discharge line
7) . Compensation Column.
8) . Shut-off valves.
9) . Pressure regulator.
10) Flow meter.
11). Overpass.
12). Flowmeter.
13). Pressure regulator.
14). Shut-off valves.

The water discharge line is required at the time of installation. When the pump section is immersed in a water body, a water flow is immediately created in the water line. At the same time, the shut-off valves on the main line are closed, while the process line is open and discharges all the pumped water. When the installation of all equipment is complete, the main line is opened and only then the process line is closed. In the same way it is used when the equipment is taken out for repair. All diagrams and sketches of the proposed installation are

collected at the end of this paper.
The guide grate of sketch #13, sketch #14, sketch #15, and sketch #16 is a welded structure made of long steel bars. For stability, the bottom has a larger area than the shaft itself. The top cover, sketch No. 16, is made in two halves and closing on the lock in the form of a tooth welded on the water pipe of the piston, sketch No. 12, which is a steel ring with a diameter of 360 mm. and thickness of 50 mm. forms a single unit with the fixed piston, thus taking all the breaking loads on itself. This closed construction allows the use of large applied forces, where force multiplied by displacement is work, and therefore energy. The enormous forces, in the form of the weight of the moving structure and Archimedes' force, extract energy far more than in any analogue in terms of cost per kW. In order to prevent detachment from the ground, the weight of the guide grid together with the weight of the piston must be much greater than the resulting pressure force, otherwise weight must be added in the form of concrete blocks at the bottom of the guide grid sketch No.23.
Let us consider the principle of operation of the proposed pumping unit on the basis of schemes No.1 and No.2 :
A certain buoyant body (pontoon) (1) together with crests and troughs of sea waves performs alternate movement strictly in vertical plane around a fixed piston, being inside a guiding grid (2). Having a certain weight, which is the sum of weights consisting of its own weight and the weight of ballast water filled in a specially designed cavity in this pontoon, as well as a certain buoyancy due to the air space inside the pontoon exerts significant pressure on the water located in chambers A and B, (scheme № 2) squeezing it alternately from the chambers A and B in the conduit. Then the water flows through the conduit to the compensation column. On the filling conduit there is an outlet process line, a shut-off valve (3) and a check valve (4).On the body of the compensation column (6) there is a pressure gauge (5). An air cushion in the upper part of the compensation column allows to partially smooth out the pulsating character of the water flow. At the outlet of the compensation column there are two water conduits, one of which is designed to supply water to the blades of the hydraulic turbine, and the other is designed to discharge excess water back into the reservoir. Each of these conduits has shut-off valves (7) and (13) pressure regulators (8) and (12) and flow meters (9) and (11), as well as manometers (5a) and (5c). Shut-off valves and flow meters are standard equipment and are included in the catalogues of many companies. Pressure regulators for these parameters should be manufactured according to the attached sketches. The general view of the complete pressure regulator is shown in sketch No. 19. The detailing is shown on the general series of sketches No.

19; No. 20; No. 21. The pressure regulator shall be made of stainless steel and shall consist of:

1) . Pressure regulator housing sketch No. 19a
2) . Passing sleeve sketch No. 19c
3) . Spacer bushing sketch #19n
4) . Pressure cover sketch no. 19m
5) . O-ring washers sketch No. 19F
6) . Spring washer sketch No. 21.
7) . Spring
8) . Spring cup sketch No. 20
9) . M38 nut.
10). Seal housing sketch no. 19c
11). Rubber rings F8mm
12). Valve sketch no. 19d and ring sketch no. 19d1
13). Rubber sleeve seal sketch No. 19E

The cross-section in the bushing No. 2 at fully open valve No. 12 is selected in such a way that the total area of the passage openings is not less than the cross-section of the water pipe F200mm.

Initially, with the spring fully released, little or no pressure is created inside the compensation column, as the spring is pressed down, the pressure inside the column begins to increase and some of the water is discharged through the conduit, pushing the valve away. As the spring is compressed, the necessary water pressure can be adjusted to pressurise the hydraulic turbine blades. At the same time, by compressing the springs on both regulators, when the required pressure is reached, spring tension must be added to the water discharge conduit regulator. In this way, both regulators will be in working condition.

The assembled scheme of this installation will allow to confirm or deny the calculated part of the project. By the pressure gauge it is possible to determine the pressure actually achieved, and by the flow meters the amount of water that will reach the blades of the hydro turbine and the amount of water that will be discharged into the water body.

Reducing the geometrical dimensions of this unit to a minimum in order to save money is not expedient, as on the same principle works a lot of analogues, which have already confirmed the performance of marine power plants. The adopted dimensions of the pumping section are optimally minimal, which is confirmed by the absence of ballast and the approximate equality of upward and downward forces. This pilot plant is intended to reveal the possibility of obtaining more energy with minimal input resources. For this purpose, large (in quantitative terms) pumping units should be used.

The cost of resources per unit of power generated, compared to alternative analogues and conventional power sources, will allow the invention to be evaluated. The cost resources per unit of power generated, compared to alternative analogues and conventional sources of energy, will allow an evaluation of this invention. The force due to buoyancy is the greater the greater the geometric dimensions of the pontoon chamber. Accordingly, the same, allows the dead weight of the pontoon section to be supplemented with ballast water. But increasing the dimensions of the pumping section also has its limit. The horizontal dimension should not be more than half the distance between two wave crests. The vertical dimension can be larger depending on the depth of the guide grid.

The buoyancy is equal to the internal air volume of the pontoon. The working pontoon in this variant is a welded structure made of ship sheet iron, geometrically resembling a parallelepiped with dimensions 2x2x4m. On the body of the pontoon sketch № 1 and № 2 there are recesses for installation of support rollers sketch № 5. In the middle and lower parts of the pontoon there are rectangular holes for installation of inlet valves sketch № 8, the upper inlet valves must be located below the waterline of the pontoon. In the inner part of the pontoon is a working cylinder with a diameter of 1.2m and height of 4m, for rigidity connected to the outer walls of the stiffeners sketch #2a. The cylinder is welded from the same ship steel. In working condition, being afloat, the upper part of the working cylinder is above the sea level. To exclude the air volume from the working chamber, in the working condition, during the assembly inside the cylinder, an intermediate sleeve is inserted Sketch No. 3. The upper flange of the pontoon sketch No. 11 is made of 30 mm thick sheet steel and the bottom is made of 15 mm thick sheet steel. Valve openings are united with the working cylinder boxes sketch № 4, separating the internal air cavity of the pontoon from the water. The fixed water line, on which the piston is rigidly fixed, is covered by a cover with a special sealing design sketch № 9. The cover consists of two halves sealed with a rubber gasket. The sealing rubber in the grooves of the cover and in the grooves of the piston has the same cross-section Sketch No. 17, and in both cases is pressed against the contact surface by the operating pressure. Even if it is not possible to exclude completely the passage of water through these seals, it is unlikely that they will be significant, and they will not present any problems either, as all the water that has passed through will flow back into the sea. To determine the force with which the experimental unit will operate, it is necessary to determine the internal air volume of the pumping section and the total weight of the section itself and of the assemblies and parts attached to it, together with ballast water, if any.

PONTOON HULL BUOYANCY CALCULATION

The buoyancy (VILI) is equal to:

V n.i = V1 - V2 - V3 - V4

Where:

V 1- volume of the parallelepiped

V 2- volume of eight roller cavities

V 3- volume for the water line of the intermediate sleeve

V 4- volume of valve boxes

$V1 = S * h = 2m. * 2М. * 4М. = 16М^3$

$V2 = 0.5m * 0.3m * 0.25m * 8 = 0.3m^3$

The internal volume of the intermediate insert covering the conduit is equal to:

$V3 = nr^2 h = 3.14 * 0.12 * 1.5m = 0.0471m^3$

The volume of boxes is calculated according to sketch No.4

$V4 = 0.2m * 0.31m + 0.2m * 0.27m + 0.6m * 0.27m : 2 == 0.271m^3$

$V = 16m^3 - 0.3m^3 - 0.047m^3 - 0.272m^3 = 15.385m^3$ So, the air volume of the pontoon section is $15.385m.^3$

Pumps section capacity calculation in 0.3m waves

The volume of chamber A is equal to: (scheme2)

$V_A = nr^2 h = 3.14 * 0.6^2 m * 0.3m = 0.339m^3$

Where r is the radius of the slave cylinder

h- accepted wave height

The volume of chamber B is equal to:

$V_B = nr^2 h - {}_{nri\,h}{}^2$

ri - diameter of the water pipe

$V_B = 3.14 * 0.6^2 m * 0.3m - 3.14 * 0.1m^2 * 0.3m = 0.{}_{33m3}$

The total volume of both chambers will be equal:

$V_{s=} V_A + V_B = 0.339m^3 + 0.33m^3 = 0.669m^3$

If we assume the period between wave crests to be 6 seconds, (which is typical for the Caspian Sea 5-6 seconds), then the capacity of the pumping section will be equal to:

V_s:6cek = $0.669m^3$: 6sec =0.111m /sec^3

CALCULATION OF THE DEAD WEIGHT OF THE PONTOON SECTION PARALLELEPIPED SIDE WEIGHTS

$P_{pr} = L * H * b * 7.8\ t/m^3$

Where: L - perimeter of the base

H-height

b - thickness of ship's iron

$7.8\ t/m^3$ - specific gravity of iron

$R_{pr} = 2{,}02m * 4m * 4m * 0{,}01m * 7{,}8\ t/m^3 = 2{,}521t$

Material needed:
Steel sheet 10mm. $S = 32.32m^2$
P = 2.521t.
THE WEIGHT OF THE BOTTOM WILL BE EQUAL TO:
$P_{dn} = S * b * 7.8t/m^3$
Where: S - bottom area
b - bottom wall thickness
$7.8t/m^3$ - specific gravity of steel
$S = S1 - s2$
Where: s1 - total area of the slab (2m*2m)
s2 - area of recesses for roller installation
b- bottom wall thickness $7,8t/m^3$ - specific weight of steel
$S1 = 2m * 2m = 4m^2$
$S2 = 0.5m * 0.21m * 4 = 0.42m^2$
$S = 4m^2 - 0.42m^2 = 3.58m^2$
$P_{dn} = P5 = 3.58m^2 * 0.015m * 7.8t/m^3 = 0.419t$
Material needed:
Steel sheet 15mm. $S= 4m^2$ P = 0,419t.
The weight of the internal stiffeners (sketch #2a) is equal to:
$R_{vnr} = \{(4m * 0.38m) - (0.25m * 0.3m * 2)\} *4 * 0.01m * 7.8t/m^3 = 0.427t$
Material needed:
Steel sheet 10mm $S=0,6m^2$ P = 0,427t.
Weight of the inner working cylinder is equal to: $R_{vz} = 2n\pm b7,8t/m^3$
Where: 2pg is the circumference length of the slave cylinder.
h - height of the slave cylinder
b - cylinder wall thickness
$7.8t/m^3$ - specific weight of steel
$R_{vz} = 2 * 3.14 * 0.6m * 4m * 0.01m * 7.8t/m^3 = 1.176t$
Material needed:
Sheet steel 10mm $S=15.072m^2$ P = 1.176t.
WEIGHT OF THE SIDE WALLS IN THE ROLLER RECESSES
sketch #2:
$P5 = \{(0.21m * 0.3m * 16) + (0.5m * 0.21m * 8)\} *0.01m * 7.8t/m^3 =0.144t$
Required material:
Steel sheet 10mm $S=1.848m^2$ P = 0.144t.
WEIGHT OF THE UPPER FLANGE OF THE PUMP SECTION
sketch #6 equals:
$R_f = R1 - R2 - R3 - R4$
Where: P1 - total weight of the slab (2,17m * 2,17m)

P2 - weight of a circle with radius 0.641m and thickness 0.015m
P3 - weight of a circle with radius 0.6m and thickness 0.015m
P4 - weight of notches in the slab, covering the channel on which the rollers move
P1 = 2.17m * 2.17m * 0.03m * 7.8t/m^3 = 1.1t
P2 = 3.14 * 0.641^2 m * 0.015m * 7.8t/m^3 =0.151t
P_3 = lg117,8t/m^3
Where: r - radius of the bore for the slave cylinder equal to 0.6m
h - boring depth equal to 0.015m
7.8t/m^3 - specific gravity of steel
P3 = 3.14 * 0.6^2 m * 0.015m * 7.8t/m^3 =0.132t
P4 = 0.16m * 0.085m *4 * 0.03m * 7.8t/m^3 =0.073t
Thus, the weight of the upper flange will be equal to: Rvf= 1.1t - 0.151t - 0.132t - 0.07t = 0.746t.
Material needed:
Sheet steel 30mm S=4,7M^2 P =1,1t.
Calculation of the weight of the support roller sketch No.5
P = P1 + P2 + P3 + P4
Where: P1 - weight of the roller base plate
P2 - weight of side support walls
P3 - weight of the roller axle
P4 - weight of the roller
P1 = 0.25m * 0.25m * 0.01m * 7.8t/m^3 =0.00487t
The area of the side stand is half the area of a circle with radius 0.04m and the area of a trapezoid with sides 0.08m and 0.18m
P2 = {pg2 : 2 +(0.04m + 0.18m) : 2 * 0,2M} * 2 * 0.05m * 7.8t/m^3 =0.0191t
P_3 = nr^2 h * 7.8t/m^3 = 3.14 * 0.022m * 0.2m * 7.8t/m^3 = 0.002t.
P4 = (nri^2 h - nr2^2 h) 7.8t/m^3
Where: r1 - radius of the roller
r2 - radius of the inner axle hole h - width of the roller
P4 = (3.14 * 0.1^2 m * 0.07m - 3.14 * 0.02^2 m * 0.07m) * 7.8t/m^3 =0.0164t Thus, the weight of all eight rollers will be equal:
P = (P1 +P2 +P3 +P4) * 8 = (0.00487t + 0.0191t + 0.002t + 0.0164t) * 8 0.339t.
Material needed:
Steel sheet 10mm. S = 0.5m^2 P = 0.03896t.
Steel sheet 50mm. S =0.362m^2 P = 0.152t.
Rolled steel F30mm L = 2m. P = 0.011t.
Rolled steel F220mm L= 1m. R= 0,296t.

WEIGHT OF FLANTS FOR INPUT VALVE FASTENING Sketch No. 7

R_f= (0.34m * 0.39m - 0.31m * 0.25m) *8 * 0.04 * 7.8t/m^3 =0.138t.

Materials needed:

Steel sheet 50mm. S = 1.06m^2 P = 0.413t.

WEIGHT OF INPUT VALVES sketch No. 8

R_k= P1 +P2 +P3 +P4 +P5

Where: P_i - weight of the flange

P2 - weight of side walls

P3 - weight of the support plane

P4 - weight of valve cover

P5 - weight of the bracket

P_i= (0.39m * 0.34m - 0.21m * 0.2m) * 0.01m * 7.8t/m^3 = 0.0072t

P2 = 0.2m * 4 * 0.06m * 0.015m * 7.8t/m^3 =0.0056t.

P3 = (0.2m * 0.2m - 0.16m * 0.16m) * 0.015m * 7.8t/m^3 =0.002t

P4 = 0.24m. * 0,24м. * 0,015м. * 7.8t/m^3 = 0.007t.

P5 - approximately equal to 0.004t.

Rcl = 0.0072t. + 0.0056t. + 0.002t. + 0.007t. + 0.004t. =

=0,0258т.

Weight of eight valves material sheet steel 15mm. S = 0.819m^2 . P = 0.096t.

Calculation of the weight of the intermediate bushing sketch No. 3

R_{pv}= P_i+ P2 + P_z+ P4

Where: P_i - weight of the outer wall

P2 - weight of the inner wall

P3 - weight of the bottom

P4 - weight of the upper flange of the intermediate sleeve

P_i= 2nri *h *b*7.8t/m^3

Where: r_i - radius of the outer surface of the intermediate sleeve

h, - height of intermediate sleeve

b, - wall thickness

7.8t/m^3 - specific weight of steel

P_i= 2 * 3.14 * 0.6m * 1.27m * 0.001m * 7.8t/m^3 = 0.373t.

P2 = 2nr2 h_i b 7.8t/m^3

Where: r2 - radius of the inner tube of the intermediate sleeve

h, -height of the inner pipe

b, - metal thickness

7.8t/m^3 - specific weight of steel

P2 = 2 * 3.14 * 0.112m * 1.295m * 0.01m * 7.8t/m^3 =0.071t.

P3 = (nri^2 bi - $nr2^2$ bi) b 7.8t/m^3

Where: r_i - outer radius of the bottom

r2 - inner circle radius bi - bottom wall thickness $7,8t/m^3$ - specific weight of steel

$P3 = (3.14 * 0.5982m * 0.015m - 3.i4 * 0.i\ 12^2 M * 0.015m) *7.8t/m^3 = 0.126t.$

$P4 = (nr32h3 - nr42h4 - nr52hs) * 7.8t/m^3$

Where: r_3 - outer radius of the upper flange of the intermediate bushing

h3 - metal thickness of the upper flange r4 - radius of the inner circle = 0,122m.

h4 - height of the inner circle =0,01m.

r5 - radius of the upper notch = 0,135m

h5 - thickness of the upper notch =0,005m.

$P4 = (3.14 * 0.64^2 m * 0.015m - 3.14 * 0.122^2 m * 0.01m - 3.14 * 0.135^2 m *0.005m) *7.8t/m^3 =0.142t$

Rpv = 0.373t + 0.071t + 0.126t + 0.142t = 0.712t.

Materials needed:

Steel sheet 10mm. $S=5,696M^2$ P=0,444t.

Steel sheet 15mm. $S=2,409m^2$ P=0,282t.

Calculating the weight of the cover sketch No. 9

Rk =P1 +P2 +P3 +P4 +P5 +P6

Where: P1 - weight of square flange

P2 - weight of the bottom ring with radii 0,23m and 0,19m r = 0,23m r = 0,11

P3 - weight of the ring r = 0,23m r = 0,11m

P4 - weight of the ring r = 0,19m r = 0,15m

P5 - weight of the ring r = 0,19m r = 0,11m

P6 - weight of ten stiffeners

P1 = P1 - P11 -P12 - P13

Where: P si - weight of square blank 2.7m * 2.7m

P11 - weight of excavation volume0,16m*0,0,06m

P12 - weight of the inner hole r = 0,11m

P13 - weight of metal volume in drilled holes 36 holes F25 mm

$P11 = 0.16m * 0.06m *4 * 0.02m * 7.8t/m^3 =0.006t.$

$P12 =3.14 * 0.11^2 m * 0.02m * 7.8t/m^3 =0.006t.$

$P13=3.14 * 0.0125^2 m * 0.02m * 7.8t/m^3 =0.003t.$

$P\ s1 = 2.7m * 2.7m * 0.015m * 7.8t/m^3 =0.551t.$

P 1 =0.551t -0.006t - 0.006t -0.003t =0.536t.

$P2 =(PG1^2 - PG2^2) *b *7.8t/m^3 =$

$= (3.14 * 0.23^2 m - 3.14 * 0.11^2 m) * 0.02m * 7.8t/m^3 = 0.02t.$

$Rz = = =(PG1^2 - pgz^2) * 0.02m * 7.8t/m^3 =$

$(3.14 * 0.23^2 m - 3.14 * 0.19^2 m) * 0.01m * 7.8t/m^3 =0.008t.$

$P4 = (3.14 * 0.19^2 m - 3.14 * 0, 15^2 m) * 0.02m * 7.8t/m =^3$

=0,006т.

$P5 = (3.14 * 0.19^2 m - 3.14 * 0.011^2 m) *0.02m * 7.8t/m^3 =$
=0,02т.
P6 =0.02t.
$R_k = 0.551t + 0.02t + 0.008t + 0.006t + 0.012t + 0.02 = =0.618t.$
Materials needed:
Steel sheet 20mm $S = 5.286m^2$ P = 0.824t.
CALCULATION OF THE WEIGHT OF WATER CONDUITS OF INTAKE VALVES
Sketch #4
$R = R_a + R_v + R_c + R_d$
Where: R_a - left sidewall area s_a
P_v -- weight of the top plane and bottom area $S_B = s_g$
P_c - weight of the right sidewall area s_c
Rd - weight of the bottom area s_d
$s_a = 0.2m. * (0.31m. - 0.05m.) + 0.05m. * 1m = 0.102m^2$.
$S_B = s_d = 0.27m. * 0.4m + 0.6m. * 0,27м. : 2 = 0.189m^2$.
$S_c = 0.2m. * (0.31m. - 0.05m.) + 0.05m * 0.4m. = 0.072m^2$.
The sum of the areas for the upper valves will be:
$S_S = s_a + S_B + s_c + s_d = 0.102m^2 . + 0.189m^2 . + 0.072m^2 . + 0.189m^2 . = 0.552m .^2$
The weight of one box of water pipe is equal to:
$P = 0.552m^2 * 0.01m. 7.8t/m^3 = 0.043t.$
The weight of the four top boxes of water conduits under the valve will be:
$R_{vk} = 0.043t. * 4 = 0,172т.$
The sum of the areas of the bottom valve boxes is equal to:
$S_Z 1 = s_a + S_B + s_c = 0.102m^2 + 0.189m^2 + 0.072m^2 = 0.363m^2$
$R_{nk} = 0.363m^2 * 0.01m * 7.8t/m^3 *4 = 0.1132t.$
The total weight of the boxes is:
$P_{E k} = R_{vk} + R_{nk} = 0.172t. + 0.1132t. = 0,2852т.$
Materials needed:
Steel sheet 10mm $S=9,28m^2$ P = 0,723t

PONTOON SECTION WEIGHT

1) . Weight of side walls2 ,521t.
2) . The weight of the bottom is 0 ,419t.
3) . Weight of internal stiffeners 0.427t.
4) . Weight of side walls0 ,144t.
5) . Weight of the slave cylinder1, 176t.
6) . Weight of the upper flange0 .746t.
7) . Weight of flanges for inlet valves 0.138t.
8) . Weight of inlet valves0 ,258t.

9) . Weight of cover0 ,618t.
10) Weight of rollers0 ,339t.
11) The weight of the intermediate bush is 0 .712t.
12) . Weight of fasteners0 ,05t.
13) . Weight of the water conduits0 ,2852t.

E =7,7812т.
Buoyancy is -15 .385m^3
Weight equals-7 .7812t.
The difference between buoyancy and weight, will give the force with which the pilot plant will operate.
F = 15.385m^3 - 7.7812t. = 7,6038т.
(1m^3 water = 1t.)
Since this force is less than the weight of the pontoon section, no ballast water will be needed to equalise the forces. Let us assume that with a force of 7.6038t the pumping section will work both when ascending to the crest of the wave and when descending from it.
Bucket hydro turbines have up to six nozzles up to 80mm in diameter. In this case, the pressure of the expiring jet will depend on the area of the jet.
Let's make some calculations: For a hydraulic unit with two nozzles, each of which is equal to 80mm. With a fully open needle, the jet pressure will be equal to:
P = F/S =7604kg/3.14 * 4^2 cm.*2 =75.6kg/cm^2 =75.6atm.
This pressure corresponds to the pressure at the dam height of 756m.
In a hydraulic unit with a single nozzle with a diameter of 80 mm, the generated pressure will be equal to:
P = F/S =7604kg/3.14 * 4^2 cm. =151.3kg/cm^2 = 151.3atm or 1513m.v.s.
In a hydraulic unit with 4 nozzles of diameter F50mm with fully open needles the jet pressure will be equal to
P = F/S =7604kg/3.14 * 2.5^2 cm*4 =96.86kg/cm2 =96.86atm. Or 968.6m.v.s.
In a hydraulic unit with two nozzles with a diameter of 50mm. with a fully open needle, the pressure of the expiring jet will be equal to: P = F/S = 7604kg/3,14*2,5^2 cm. * 2 = 192.05kg/cm^2 = 193.7atm. Or 1937 m.v.s.
In a hydraulic unit with 4 nozzles with a diameter of3.5cm. the pressure of the expiring jet will be equal to:
P = F/S = 7604kg/3.14*1.75^2 cm*4 = 197.7kg/cm^2 =197.7atm. Or 1977 m.v.s.
In a hydro unit with six nozzles with a diameter of3.5cm.
The pressure of the expiring jets will be equal:
P = F/S = 7604kg/3.14*1.75^2 cm*6 = 131.8kg/cm^2 =131.8atm. Or 1318 m.v.s.
The obtained data are the initial data for the hydroturbine manufacturer to

determine the capacity of the ordered hydraulic unit.

Power [kW] = Head [m] * Water flow rate [t/sec] * Acceleration free fall [9.81 m/sec^2] * efficiency [0.6]

W =1937m * 0.111m^3 /sec * 9.8m/sec^2 * 0.6 = 1264kW (two nozzles 50mm diameter at a wave height of 0.3m).

W = 1937m * 0.333m^3 /sec * 9.8m/sec^2 * 0.6 = 3792kW (two nozzles 50mm diameter. At a wave height of 1 metre).

W =3000m* 0.111m^3 /sec*9 .8m/sec^2 * 0.6 = 1958kW (two nozzles

35mm diameter at 0.3m wave).

W =3000m* 0.333m^3 /sec*9 .8m/sec^2 * 0.6 = 5874kW (two nozzles

35mm diameter at 1 metre wave).

To determine the stability of the entire structure, the weights of all structural elements, including fixed elements, must be determined.

Piston head weight calculation sketch No. 12

P = Pdn + Pver + P1 + P2 + P3 + P4 + P5 + P6 + P7

Where:Rdn - weight of the bottom

Top - weight of the upper plane

P1 - weight of rings F1,196m on F 1,062m.

P2 - weight of rings Φ1,142m. to Φ 1,062m.

P3 - weight of welded insert F1,1m. b=0,02m. h = 0,34m.

P4 - weight of valve box "B"

P5 - weight of valve box "C"

P6 - weight of spigots

P7 -£ weight of valves "A" "B" "C"

Rdn = nr^2 * b * 7.8t/m3 - 6 * nr1^2 * b * 7.8t/m3

Where: r - piston outer diameter

r1 - diameter of inlet valve bores

b - sheet thickness

7.8t/m^3 - specific weight of metal

Rdn = 3.14 * 0.598^2 m. * 0,02м. * 7.8t/m^3 - 6 * 3.14 * 0.0475^2 m. * 0,02м. * 7.8t/m^3 =0.1685t.

Rverh= Rdn -3,14 * 0,11^2 m. * 0,02м. * 7.8t/m^3 =0.1625t.

P1 = (3.14 * 0.598^2 m. - 3.14 * 0.531^2 m.)* 0.02m. * 7.8 t/m^3 = 0.1482t.

P2 = (3.14 * 0.571^2 m. - 3.14 * 0.531^2 m.) * 0.02m. * 7.8t/m^3 = 0.0864t.

P3 = 2*nrh * b *7.8t/m3 = 2 * 3.14 *0.55m. *0,34м. * 0.02m * 7.8t/m^3 =0.1832t.

P4 = {(3.14 * 0.5252m. - 3.14 * 0.11^2 m) * 2 + (2 * 3.14 * 0.525m. * 0.06m.) -

$(12 * 3.14 * 0.0475^2$ m.)} * 0,01м. * $7.8t/m^3 = 0.1424t.$
P5 = $\{(3.14 * 0.4^2$ m - $3.14 * 0.11^2$ m.) - $(6 * 3.14 * 0.045^2) + (3.14 * 0.8$m. * 0.15m.)} * 0.01m * $7.8t/m^3$ = 0.0625t.
P_6 = (six pipes F86 0,34m long) = 0,032t.
Calculating the weight of the return valves "A" "B" "C" sketch No. 18
P7 =($P_{11} + P_{12} + P_{13} + P_{14}$) * 18 = (0.004t. + 0.0034t. + 0.0035t. + 0.0001t.) * 18 = 0.198t.
Where: P_{11} - weight of valve body sketch No.18a
=0, 004т.
P_{12} - weight of valve plate sketch No. 18d
= 0,0034т.
P_{13} - weight of valve base sketch No. 18c
= 0,0035т.
P_{14} - weight of rubber bushing sketch #18c
=0,0001т.
The weight of the piston head will thus be equal to:
P = 0.1685t. + 0.1625t. + 0.1482t. + 0.0864t. + 0.1832t. + 0.1424t. + 0.0625t. + 0.032t. + 0.198t. =1,1837т.
Materials needed:
Steel sheet 20mm. S = $12.017m^2$ P = 1.875t.
Steel sheet 10mm. S = $2.67m^2$ P = 0.203t.
Pipe F86 L = 3m. p =
Rolled steel F200mm L = 3.5m.R = 0.857t.

The guide grid is designed to give the pump section a strictly vertical movement,
This neutralises all lateral forces of waves and currents from different wind directions. In addition, the guide grille and the fixed piston, which is united with it, absorb all the reactive forces generated by the pumping section through the closing cover, thus allowing a high pressure to be generated. The weight of the supporting structure of the grid, together with the weight of the piston and the adjoining conduit, must be much greater than the force with which the pump unit operates. If the weight of the guide grating, the piston, the water pipe and the cover is not sufficient to neutralise the force with which the pumping unit is operating, concrete blocks are added to the base of the guide grating Sketch No. 23
Calculation of the weight of the lattice guide sketch No. 13; No. 14 No. 15
The shaft of the guide grating sketch №13 is made of channel №12 in the amount of 32m + lintel 56m angle 60 x60mm.40m. The weight of one metre of channel №12 is 10.4kg/m. R_{sh} = (32m. +56m.) * 0,0104t. =0,915т.

R_u = 40m. * 0,002826т. = 0,113т.

The weight of the top flange sketch #14 is:

R_{vf} = (2,51m * 2,51m. - 2,27m. * 2,27m.) * 0,02m. * 7.8t/m3 = 0.179t.

The bases are welded from channel #12 and 20mm thick steel sheet. Sketch No. 15 and has a weight:

The total quantity of channel is equal to

R_{sh} = (3.27m. *4 +2.27m. * 4) * 0.0104t/m. = 0,2304т.

On the base are eight 0.5m x 0.5m squares, 20mm thick. And four more squares cut diagonally for stiffeners

R_k = 0.5m. * 0,5м. * 12 * 0,02м. 7.8t/m^3 = 0.468t.

The weight of the guide grille cover sketch #16 will be:

$R_k = Rf + R_z + R_r$

Where:R_f - weight of flange

P_z - weight of the lock

P_p - weight of stiffeners

R_f= (2.51m * 2.51m - 2.27m * 2.27m) * 0.02m. *7.8t/m^3 =0.179t.

R_z =($PG1^2$ - $PG2^2$) * h* 7.8t/m3

(3.14 * 0.225^2 m - 3.14 * 0.11^2 m)0.15m * 7.8t/m^3 =0.141t.

R_p =(0.28m. + 0.01m.) : 2 * 1.03m * 6 * 0.02m. * 7.8t/m^3 =0.1397t.

R_k = 0.179t. + 0.141t. + 0.1397t. = 0,4597т.

CALCULATION OF THE WEIGHT OF THE PART OF THE CONDUIT PRESSURISING THE SEA SHELF

The water line is made of thick-walled steel pipe

F219mm. And the wall thickness is 10mm. The approximate length of the conduit to the compensation column is 8m. The weight of the conduit will be equal to:

R_{vod} = 2nrbl A8T m^3

Where r is the radius of the conduit

b - thickness of the water pipe wall

L - length of the calculated section of the water pipeline

7.8t/m^3 - specific weight of steel

R_{vod}= 2 * 3.14 * 0.11m. * 0,01м. * 8м. * 7,8t/m^3 =0,431t.

The total weight of the metal structure exerting pressure on the sea shelf will be:

P = 1.1837t. + 0.915t. + 0.113t. + 0.179t. + 0.2304t. +

+ 0.468t. + 0.4597t. + 0.431t. = 3,9798т.

Materials needed:

Channel No.12 L= 116m. P = 1,206t

Angle 60 x60mm. L = 48m. P = 0,135t.

Steel sheet 20mm. S = 15.3m^2 P = 2.433t.

Steel casting F470mm x200mm. P = 0,27t.

This weight is not sufficient to compensate for the force with which the pumping section will operate(7,604t).

If two blocks of reinforced concrete are made with dimensions corresponding to the extensions of the base of the guide grid, namely: 3.2m x 0.5m x 1.3m.

Then the weight of each block will be equal:

Rbl = 3.2m * 0.5m. * 1,3м. * 2.4t/m^3 = 4.992t.

The total weight neutralising the working pressure of the pumping section will be the weight of the concrete blocks, the weight of the guide grid and all its elements:

P = 4,992. *2 + 3,98т. =13,964т.

Concrete blocks in water will be lighter by the weight of the volume of water displaced by them and this will amount to:

P1 = 3.2m * 0.5m. * 1,3м. * 2 * 1 t/m^3 = 4.16t.

Then the actual compensating force of the pontoon buoyancy will be equal to:

P = 13.964t. - 4,16т. = 9,804т.

This is 2.2 tonnes more than the force with which the pontoon will work, and is sufficient to steady the entire structure under wave loading.

For the manufacture of the expansion column and water conduits will be required:

Steel sheet 20mm. S=17,883м^2 P =2,789t.

Pipe F 219 x10мм L=10м. P= 0,3t

Flanges DU200mm in the amount of 30pcs.

Bends DU200mm x 12mm 20 pcs.

The pressure regulator, as mentioned above, is made of stainless steel, and is designed to maintain stable water pressure when it is fed to the blades of the hydraulic turbine. When this pressure is slightly exceeded, a second pressure regulator discharges the excess water back into the reservoir. The general view of the pressure regulator is shown in sketch No. 19. Valve No. 12, pressed by spring No. 7, will open only when the water pressure exceeds the pressure created by spring No. 7 and is fully pushed to the stop against the upper plane of throughput sleeve No. 2. In this case, the total cross-section of the eight holes in the bushing No. 2 will be slightly larger than the cross-section of the water pipe of 200 mm. The working pressure is selected by gradual compression of spring No.7 by nut No.9. Let's calculate the materials necessary for manufacturing the pressure regulator.

The body of the pressure regulator sketch No.19a is welded, made of 20 mm thick stainless steel sheet.

Top flange S = nr^2 and is equal to 3.14 * 0.3352m. = 0,353м^2

Side walls S = 2nrh = 2 * 3.14 * 0.245m * 0.335m =0.516m^2
Bottom S = nr^2 = 3.14 * 0.2352m= 0.174m^2
Pipes S = 2nrh = 2* 3.14 * 0.11^2 m * 0.23m * 2 = 0.035m^2
Flanges S = nr^2 = 3.14 * 0.18^2 m * 2 = 0.2m^2

S_z = 0.353m^2 + 0.516m^2 + 0.174m^2 + 0.2m^2 + 0.035m^2 = 1.278m ^{2}

two pressure regulators 2.556m^2

The bushing sketch no. 19c is made of stainless steel steel casting F 460mm H =120mm two blanks.

Spacer bushings sketch #19p can be made of steel castings F135mm L=400mm Or from sheet steel 20mm pre-welded to the required diameter well welded on the prepared deep chamfer. S = 2nrh *2 =2*3,14 * 0,07m * 017m * 2 = 0,15m^2

The pressure cover sketch no. 19m is made of 20 mm stainless steel. For two covers

S = nr^2 = 3.14 * 0.1252m * 2 = 0.1m^2

Washers for O-ring seals sketch No. 19F are made of stainless steel sheet 20 mm total area of blanks will be

S= nr^2 =3.14 * 0.072m * 4 *2 = 0.13m^2

The washer under the spring sketch №21 is made of stainless steel sheet 20 mm for two washers the area of blanks will be equal to

S= nr^2 =3.14 * 0.055m^2 * 2 = 0.02m^2

Washers for fixing the rubber seal of the pressure regulator valve sketch No. $19d_1$ are made of 20 mm stainless steel sheet.

(two pressure regulators)

S= nr^2 = 3.14 * 0.1552m. * 2 = 0,15м^2

Spring cup sketch #20 is made of 20 mm stainless steel sheet.

The lower flange is bored from 20mm stainless steel sheet.

S = nr^2 = 3.14 * 0.2252m= 0.159m^2

Stiffening ribs made of stainless steel sheet 20 mm.

S = (0.45m - 0.25m) * 0.2m + (0.45m - 0.25m) * 0.35m : 2 * 4 = 0.14m^2

The cup is welded, welded and then bored from 20mm stainless steel sheet.

S =2nrh = 2 *3.14 * 0.0725m * *0.35m = 0.16m^2

The upper plane of the cup is made of 20 mm stainless steel sheet.

S= nr^2 =3.14 * 0.07^2 m =0.015m^2

S_z = 0.159m^2 + 0.14m^2 +0.16m^2 + 0.015m^2 = 0.474m^2

For two regulators S^ = 0.474m^2 * 2 = 0.948m^2

Seal housing sketch #19c welded (argon welding), made of stainless steel

castings f450mm * *300mm 2pcs and stainless steel sheet 20mm.
$S = \pi r^2 = 3.14 * 0.3252m * 2 = 0.664m^2$
Pressure regulator valve sketch No. 19d welded, made of individual blanks: steel n/r rolled F60mm. L =2m.
And stainless steel sheet 20mm $S = \pi r^2 = 3,14 * 0,16^2 m * 4 = 0,33m^2$. The rigid rubber sealing collar sketch no. 19E is fixed by the pressure ring sketch no. 19d1 using six countersunk M6 stainless steel bolts.
Below is the table of material consumption for the production of the pump section. Based on the summarised data shown in this table, a table of the cost of materials required in local currency and in dollar equivalent, as of the end of 2016, will be drawn up below.

	TABLE **No.1** Cost of Necessary Materials	If honestly Required material	**Price per Unit Man.**	Price per Unit $	Total **Price Man.**	Total **price** S	**Note**
1	Steel plate 10mm	69.046 m² 5.759 tonnes	1257,8	704,1	11320,2	6339,1	9 leaves 10* 1500'6000mm
2	Steel sheet 1Bm.m	7,228 и¹ 0,797 т	1213,6	679,6	2127,2	1359,2	2 sheets **15*** 1500* SOOOmm
3	Steel sheet 20mm	50.486 M¹ 7,921 T	1222,6	684,7	8558.2	4792,6	7 sheets 20* 1500* 6000mm.
4	Steel sheet 30mm	6г 7 m² 1,111t	1263,7	707,7	1263,7	707,7	**1 sheet** 30*1500' 6000mm.
5	Sheet steel B om.m	1.122 m¹ 0.565 t	1355,3	758,9	1355,3	758,9	1 sheet B0* 1500* 6000mm.
Б	Rolled steel FZOmm	2 м 0,011 т	1462.7 **Man/t**	819,2 $A	16,1	9,0	
7	Rolled steel F200mm	3,5 м 0,857 т	1462.7 **Man/t**	819,2 S/T	1253,5	702,1	
В	Rolled steel F220mm	1 м 0,296 т	1462,7 Man/t	819,2 SA	432,9	242.5	
9	Pipe F 86 x 6 mm	2L m	11.5 Man/metre	6.44 5/metre.	28.75	16.1	
ю	Pipe F219x 10mm	18 м	1134,1	635,3	310,3	190,6	0,300 т
11	Channel No. 12	0.5 т	1296,1	725,8	648,1	362,9	
12	Corner 60 x AND mm	0,3 т	1296,1	725,8	388.8	217,7	
13	Steel sheet 20mm	5,048m.²	5745,1	3217,4	5745,1	3217,1	**1 sheet** 20*1500* 6000mm
14	Stainless steel castings F 170mm H 200mm	2 nt.	4000.0	2240.0	2000.0	1120,0	0,5т
15	Steel n/a CAST steel F150 H 300 mm	2 pcs.	4000,0	2240,0	2800,0	1568,0	0,7т
16	Stainless steel moulding F460 H 120mm	2 pcs.	4000.0	2240.0	1100,0	784,0	0,35т
17	Steel n/r CAST F 135mm 400m	2 pcs.	4000,0	2200,0	400,0	220,0	0,1т

	m						
					40328,45	22608,1	

		Stainless steel sheet 20mm	
Pressure regulator housing	sketch #19	2,556	$м^2$
Prostate bushing	№19n	0,15	$м^2$
Pressure cover sketch	№19m	0,1	$м^2$
Seal washer sketch	№19F	0,13	$м^2$
		5,048	$м^2$
Spring washers sketch	№21	0,02	$м^2$
Washers for mounting of threaded bracket.	sketch #19di	0,15	$м^2$
Spring tumbler sketch	№20	0,948	$м^2$
Seal housing sketch	№19c	0,664	$м^2$
Valve R.D. sketch	№19d	0,33	$м^2$

	Table No.2 Costs of materials	Quantity of Materials Required	Price per Unit Man.	Price per Unit $	Sum of Price mans	Amount Price $	Note
1	F30mm rebar	0,8 т	7,6	4,25	980,4	548,3	1m-0.0062t
2	Concrete	$12м^3$	191,5	107,2	2298,1	1286,8	
3	Flanges DU200mm	40 pcs.	55,98	31,35	2239,2	1253,9	
4	Shut-off Fittings DU200	4 pcs.	1729,4	968,5	6917,6	3873,8	
5	Reverse Valve DN200	1 pc	1915,1	1072,5	1915,1	1072,5	
6	Flowmeter DN 200mm	2 pcs.	4615,3	2584,6	9230,5	5169,1	
7	250atm pressure gauge	3 pcs.	107,0	59,92	321,0	179,76	
8	Cable el.	150 м	-	-	-	-	
9	Zinc met.	30kg	5154 man/t	2886 $/т	154,6	86,58	
10	Paint	150 kg	7,37	4,12	1104,8	619,1	
11	Electrodes	0,5 т	3.71man/kg	2,07	1856,2	1038,8	
12	Bolts with nuts M 24 x100	40 pcs.	1.4 man	0,78	56 mana.	31,36	
13	Bolts M16x45	130 pcs.	0.8 man	0,45	104 man	58,24	
14	Anchors Bolts M24x100	4pc	2.6 man	1,45	10.4man	5,8	
15	Bolts M24x70	25 pcs.	1 man		25 mana.		
16	Stainless steel countersunk screws. headM6x25	15 pcs.	0.4 man	0,22	6 mana.	3,36	
17	Bolts M16x40	20 pcs.	0.75 man	0,42	15 mana.	8,4	
18	Stainless steel bolts M22x50	20 pcs.	1.5 man	0,84	30 mana.	16,8	

19	Stainless steel bolts M12x40	20 pcs.	0.8 man	0,44	16 mana.	8,96	
20	Bends F200x12mm	20 pcs.	153,2	85,8	3063,6	1715,8	
					30343,5	16977,4	

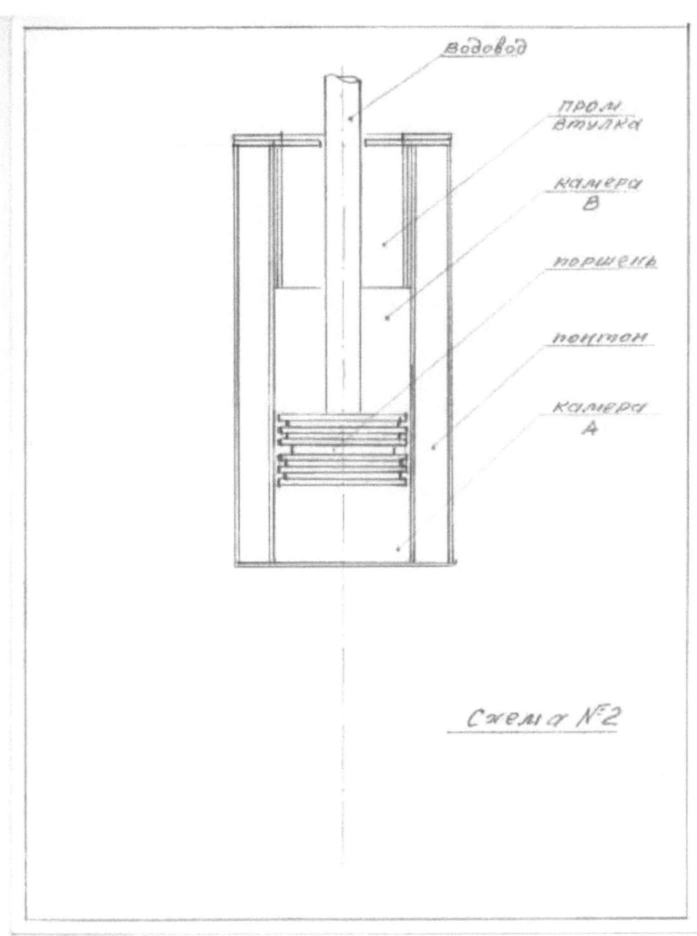

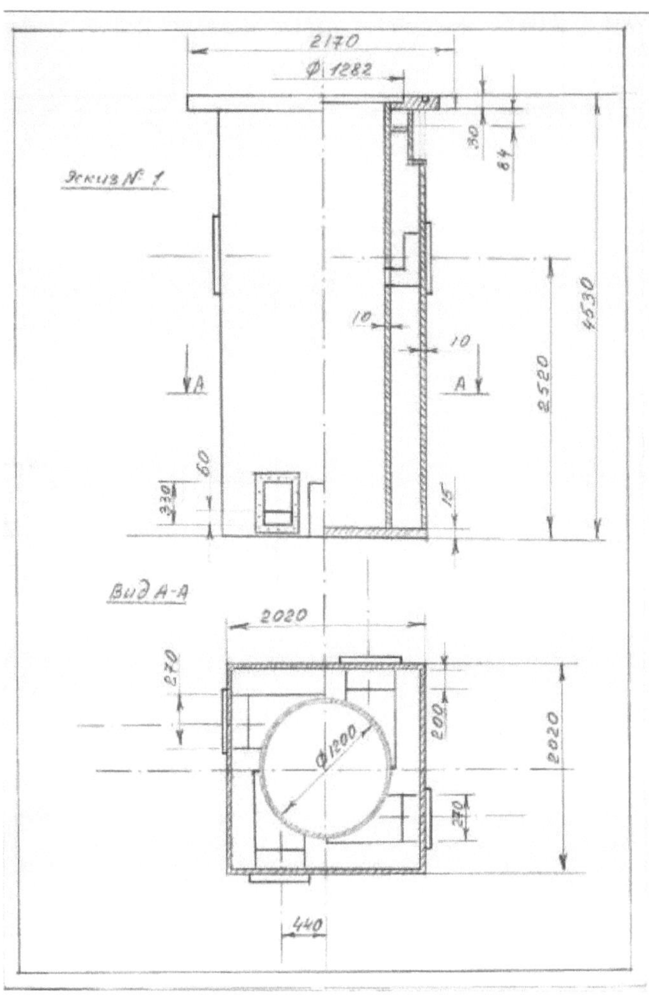

Эскиз № 1
2170
Ф 1282
30
84
4530
2520
10
10
А
А
60
330
15
Вид А-А
2020
2020
270
200
Ф1200
270
440

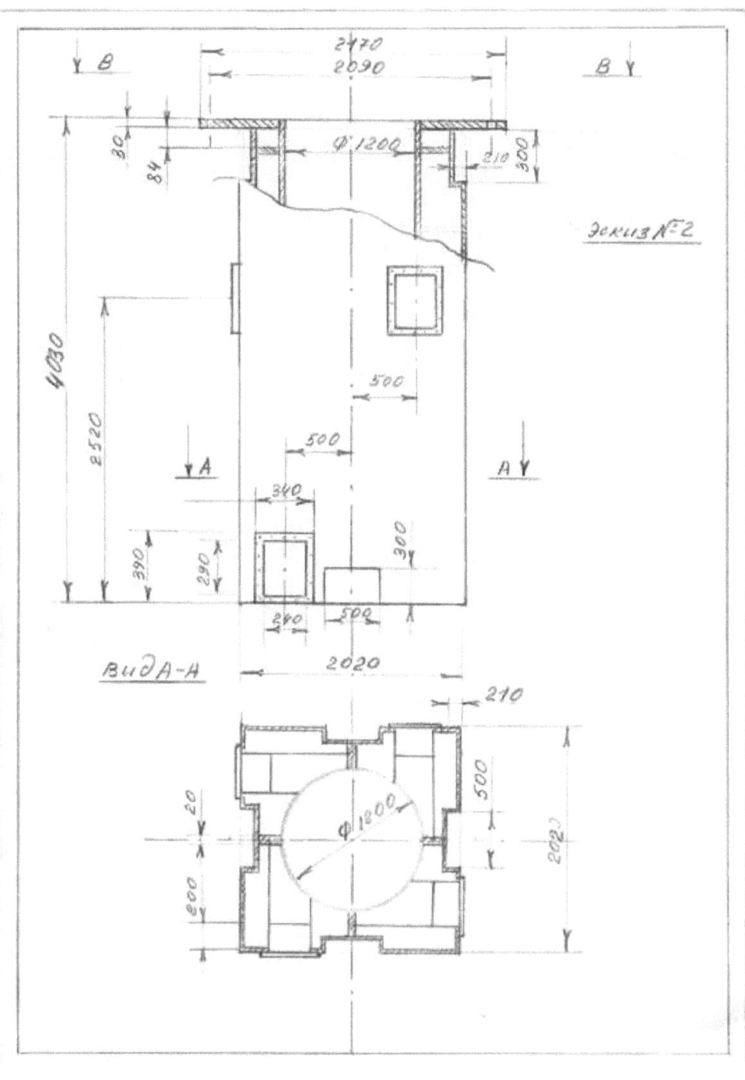

2470
2090
В
В
30
84
Ф 1200
210
300
Эскиз №2
4030
2520
500
500
А
А
340
390
290
300
240
500
Вид А-А
2020
210
500
20
Ф 1200
2020
200

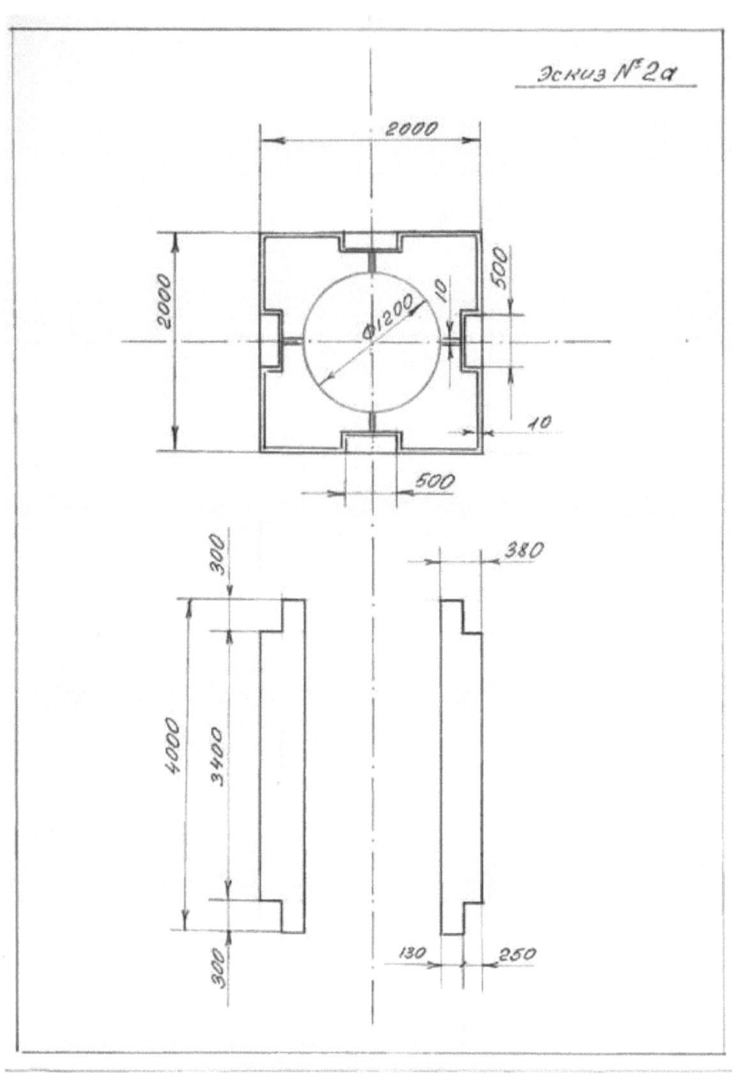
Эскиз №2а
2000
2000
500
10
Ф1200
10
500
300
380
4000
3400
300
130
250

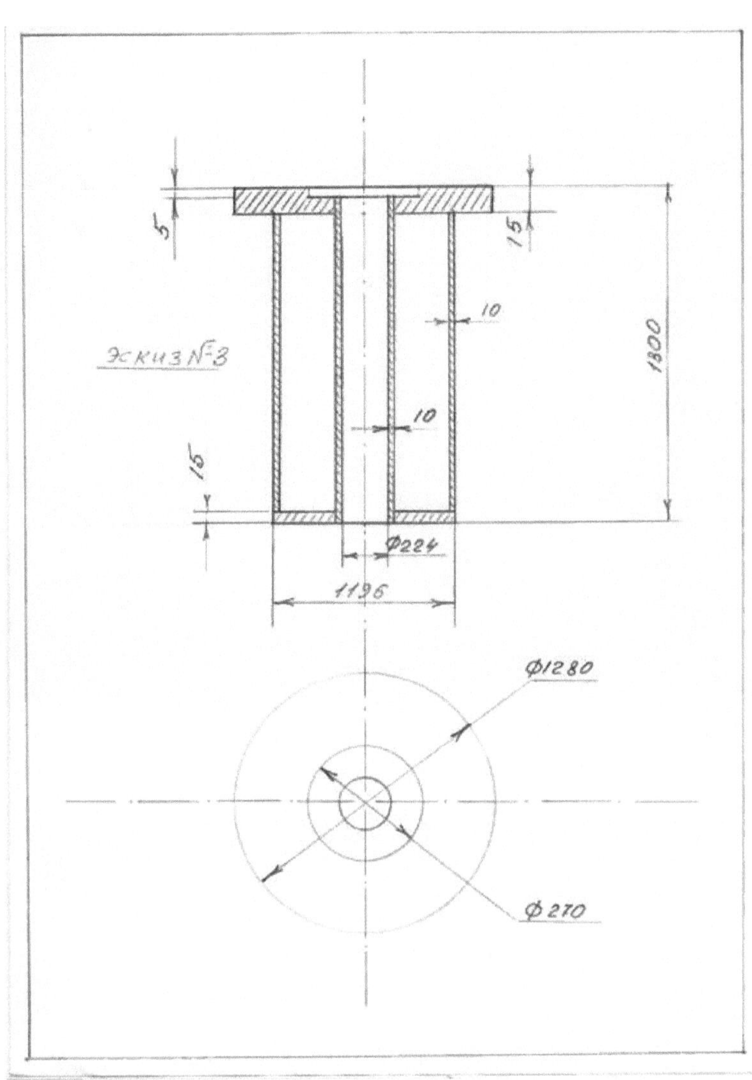
Эскиз №3
5
15
10
1300
10
15
Ф224
1196
Ф1280
Ф270

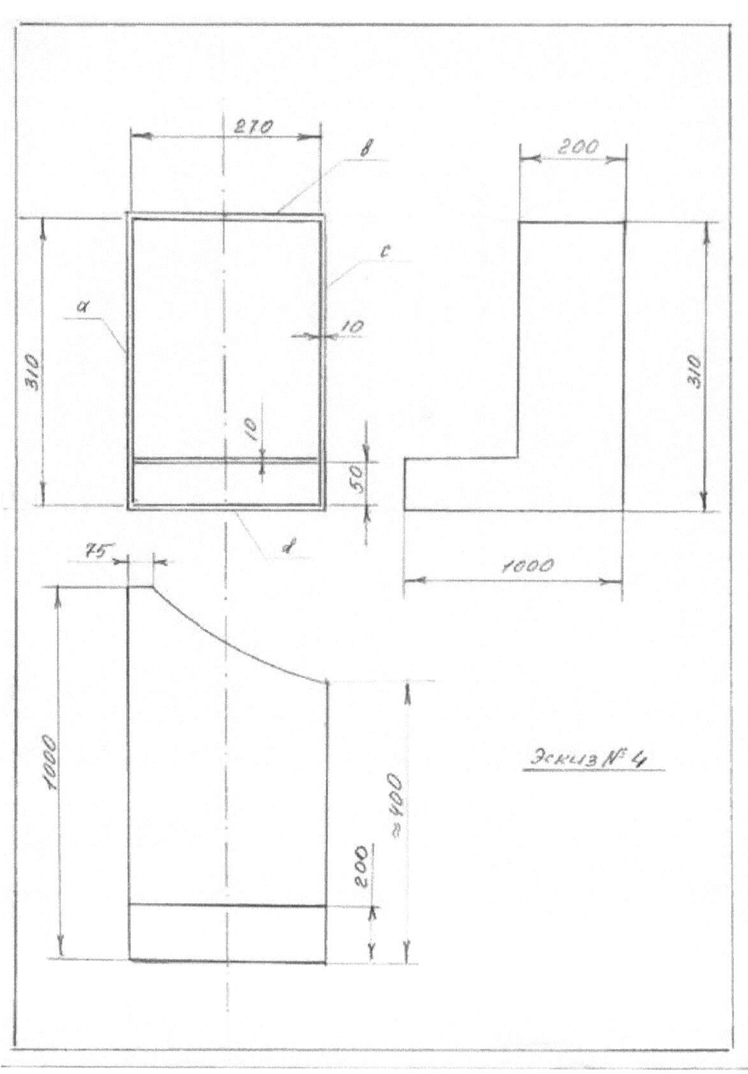
270
в
200
с
а
10
310
310
10
50
d
75
1000
1000
≈400
200
Эскиз № 4

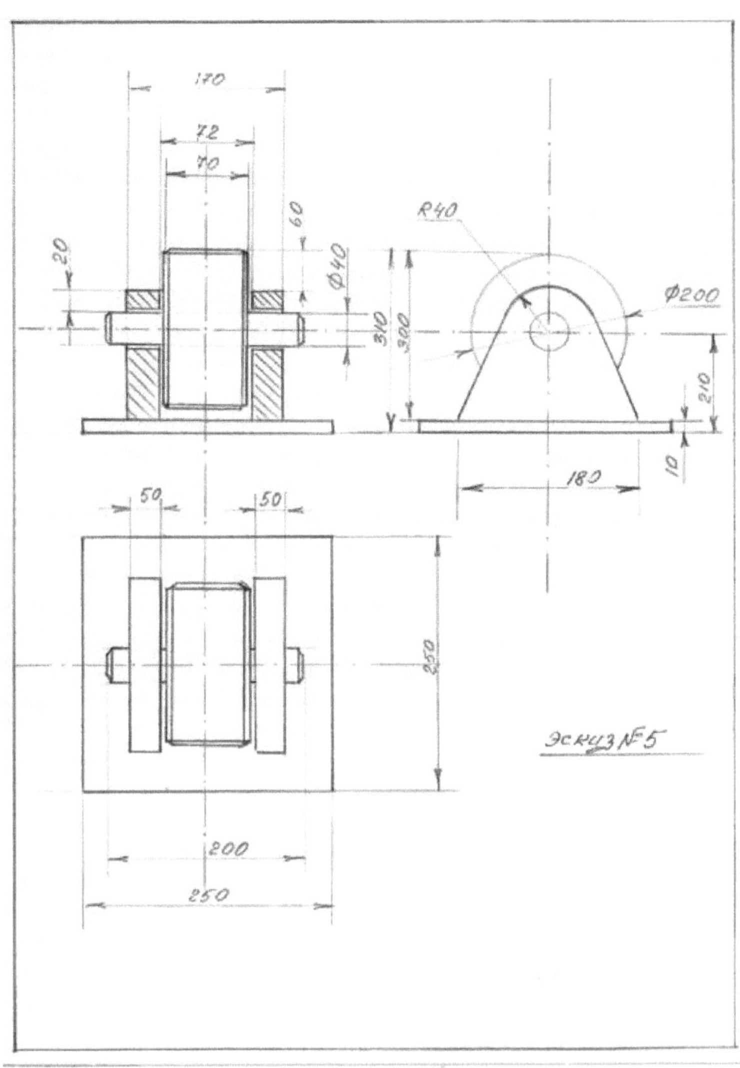

170
72
70
60
20
Ф40
310
300
R40
Ф200
210
10
180
50
50
250
200
250
Эскиз №5

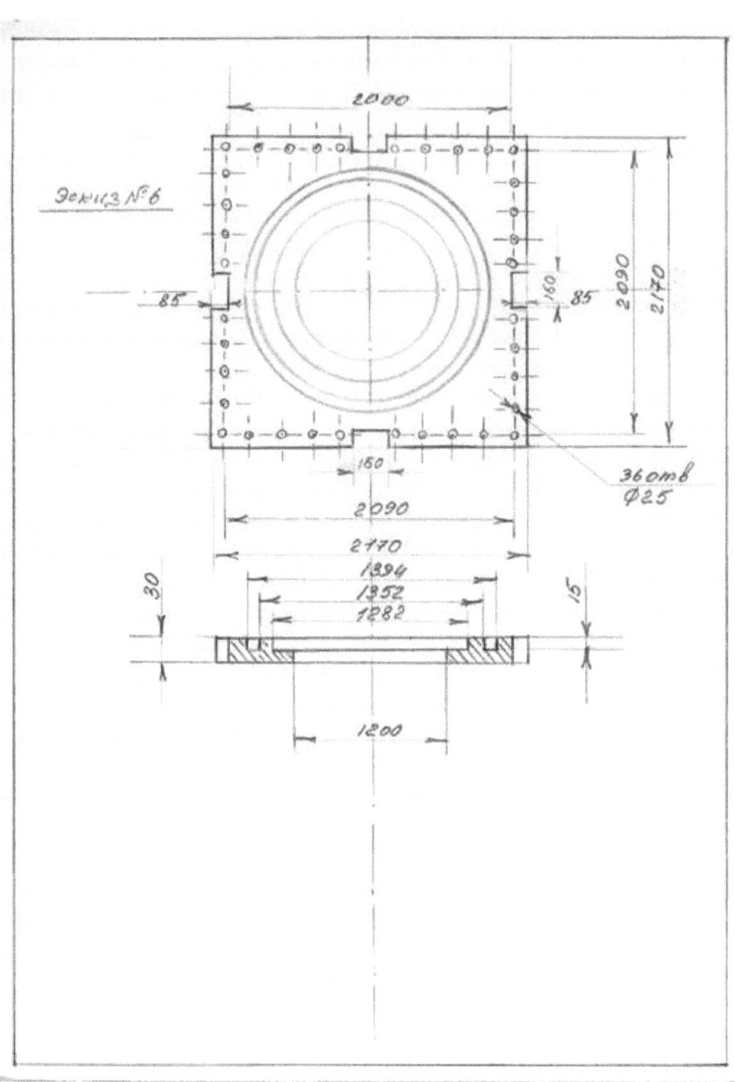
Эскиз №6
2000
2090
2170
85
85
160
160
36отв
Ø25
2090
2170
1394
1352
1282
30
15
1200

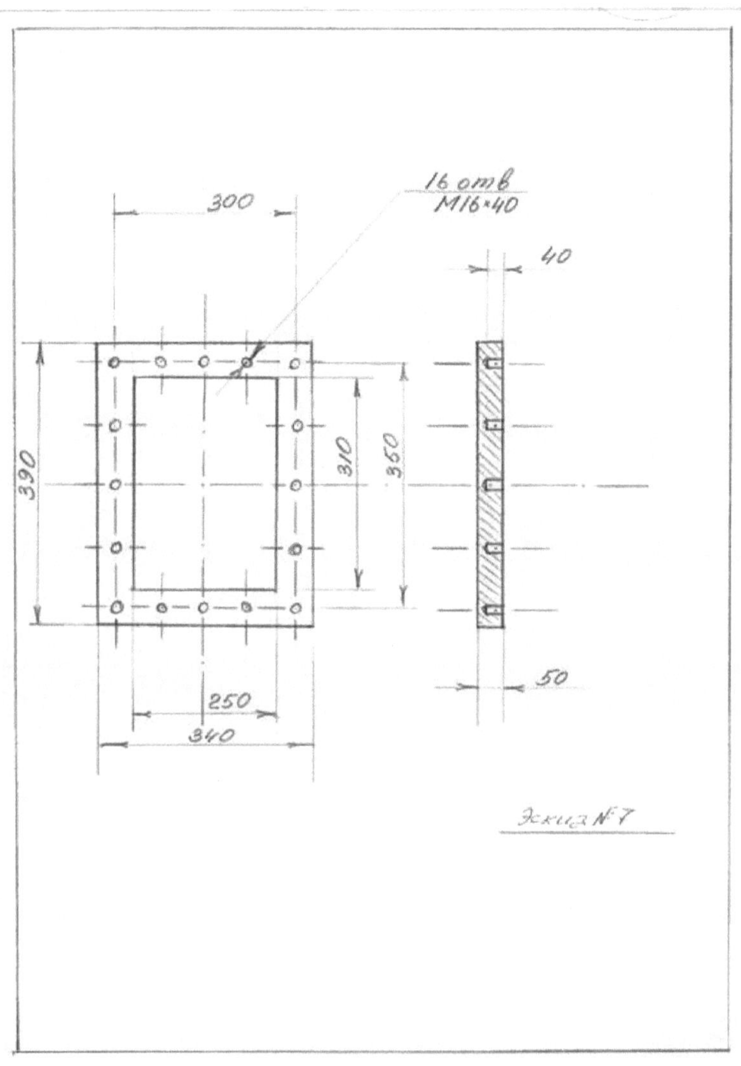
16 отв
M16×40
300
40
390
310
350
250
340
50
Эскиз №7

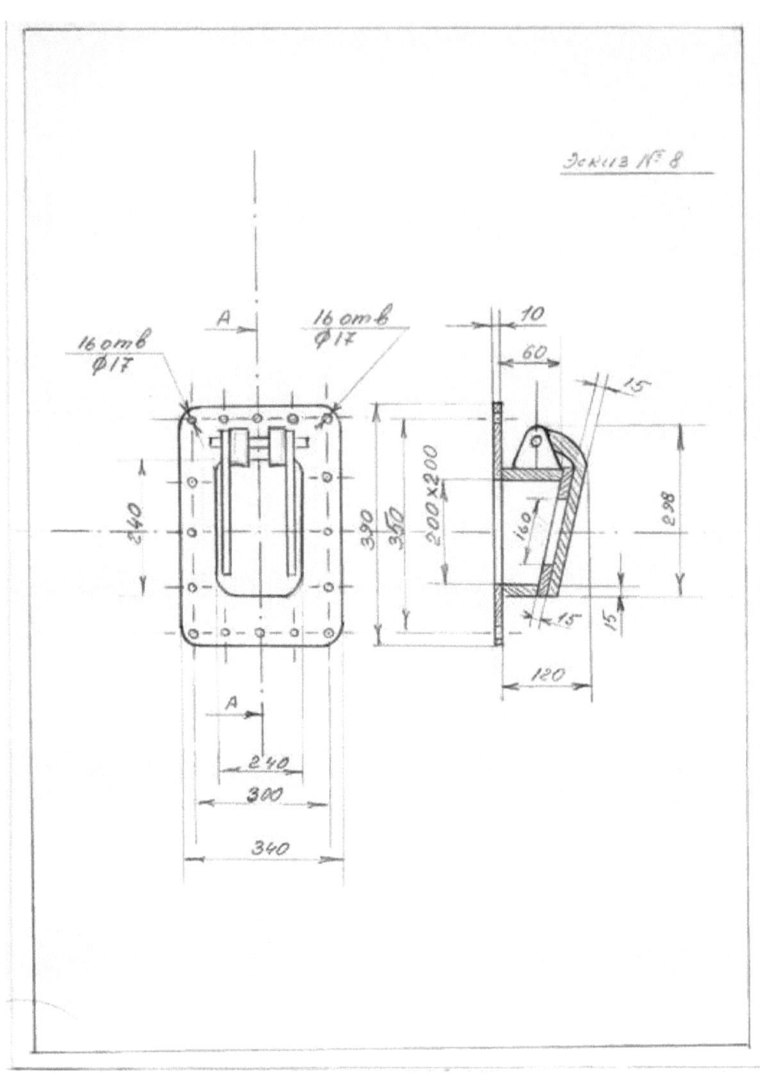

Эскиз № 8
16 отв
Ф17
16 отв
Ф17
А
А
10
60
15
240
390
350
200x200
160
298
15
15
120
240
300
340

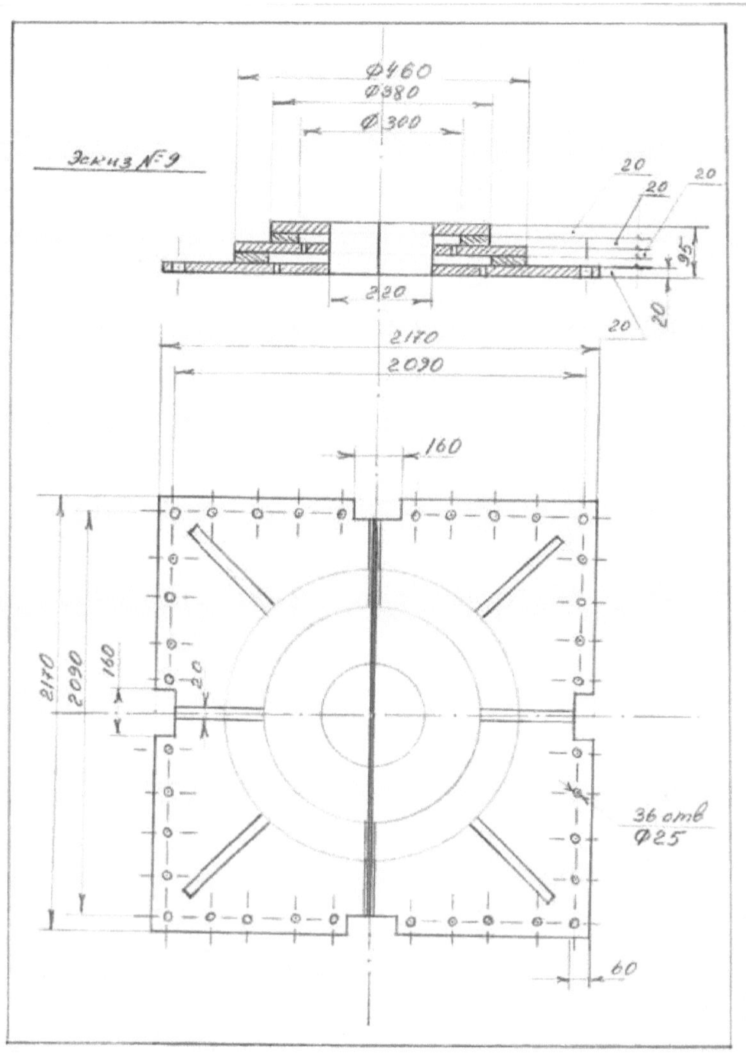

Эскиз №9
Ф460
Ф380
Ф300
20
20
20
95
220
20
20
2170
2090
160
2170
2090
160
20
36 отв
Ф25
60

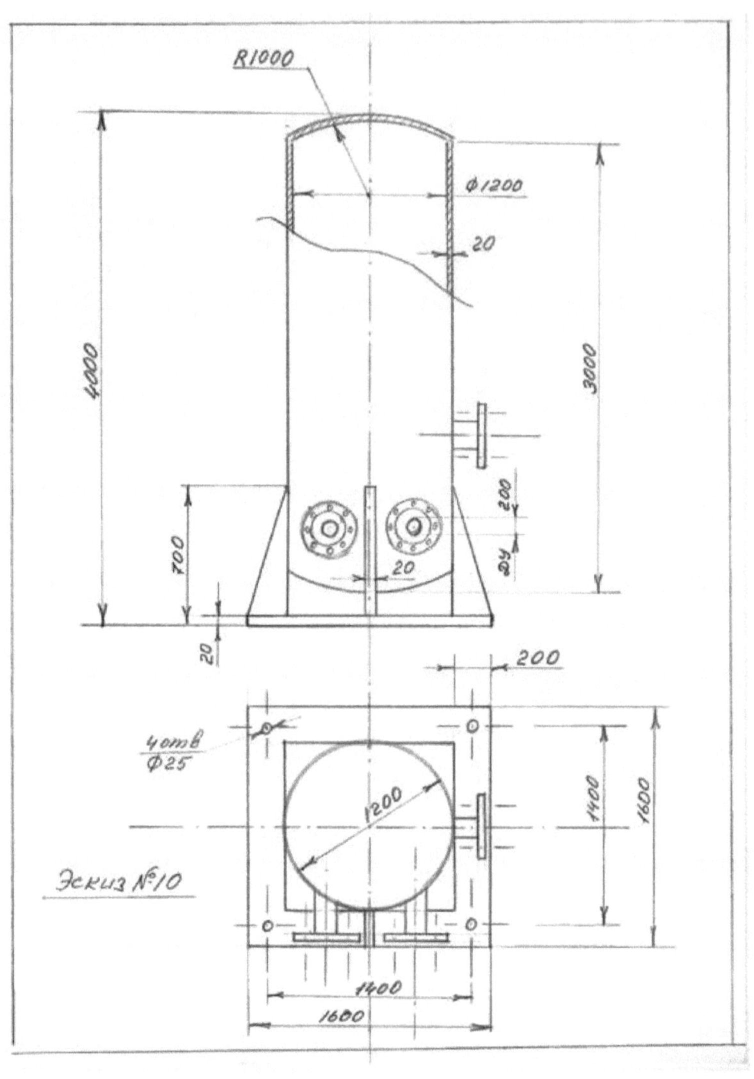
R1000
Ф1200
20
4000
3000
700
200
Ду
20
20
200
4отв
Ф25
1200
1400
1600
Эскиз №10
1400
1600

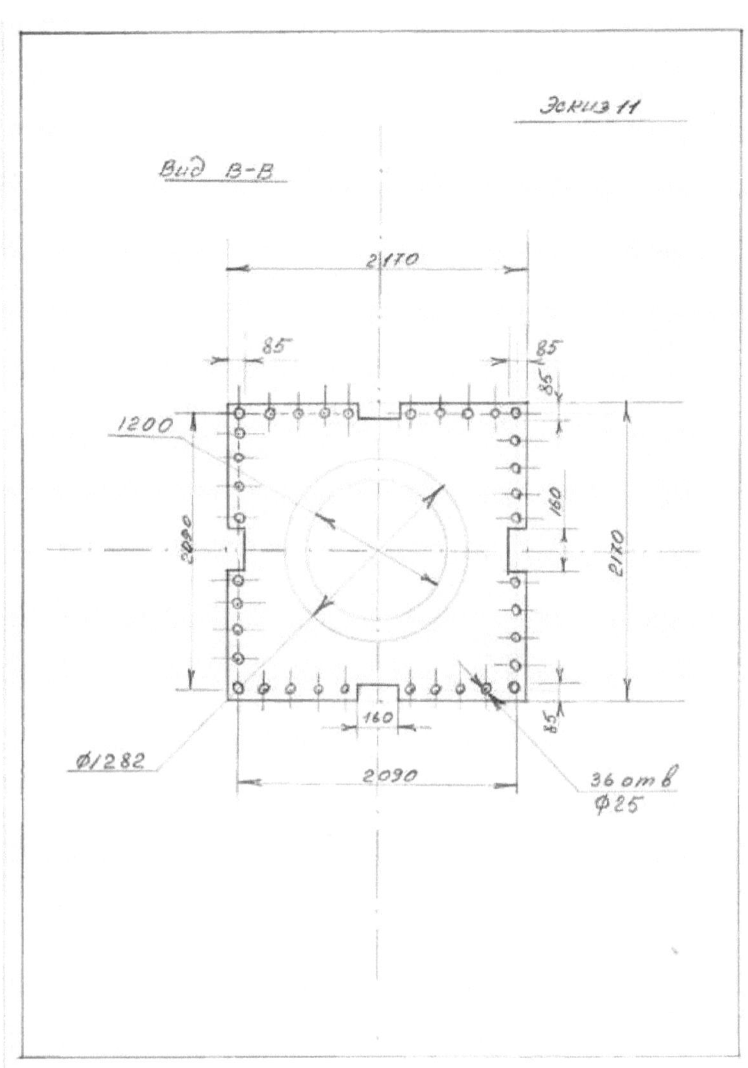
Эскиз 11
Вид В-В
2170
85
85
85
1200
160
2090
2170
Ø1282
160
85
2090
36 отв
Ø25

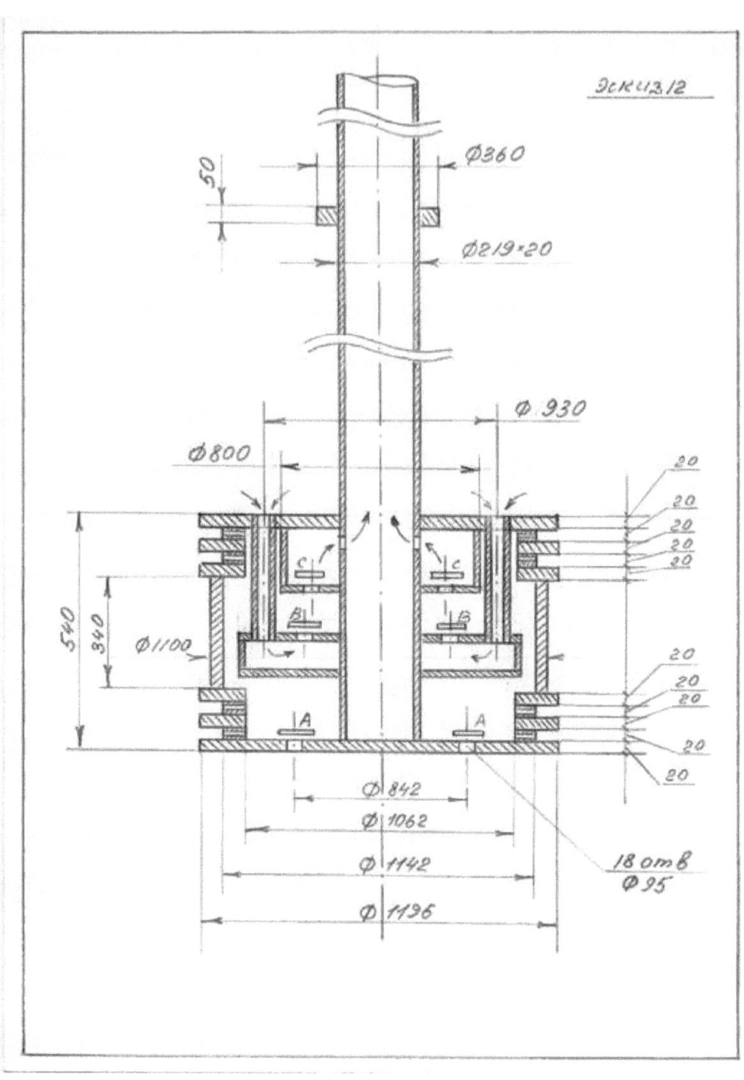
Эскиз 12
Ф360
50
Ф219×20
Ф 930
Ф800
20
20
20
20
20
C
C
B
B
540
340
Ф1100
20
20
20
20
20
A
A
Ф 842
Ф 1062
Ф 1142
Ф 1196
18 отв
Ф95

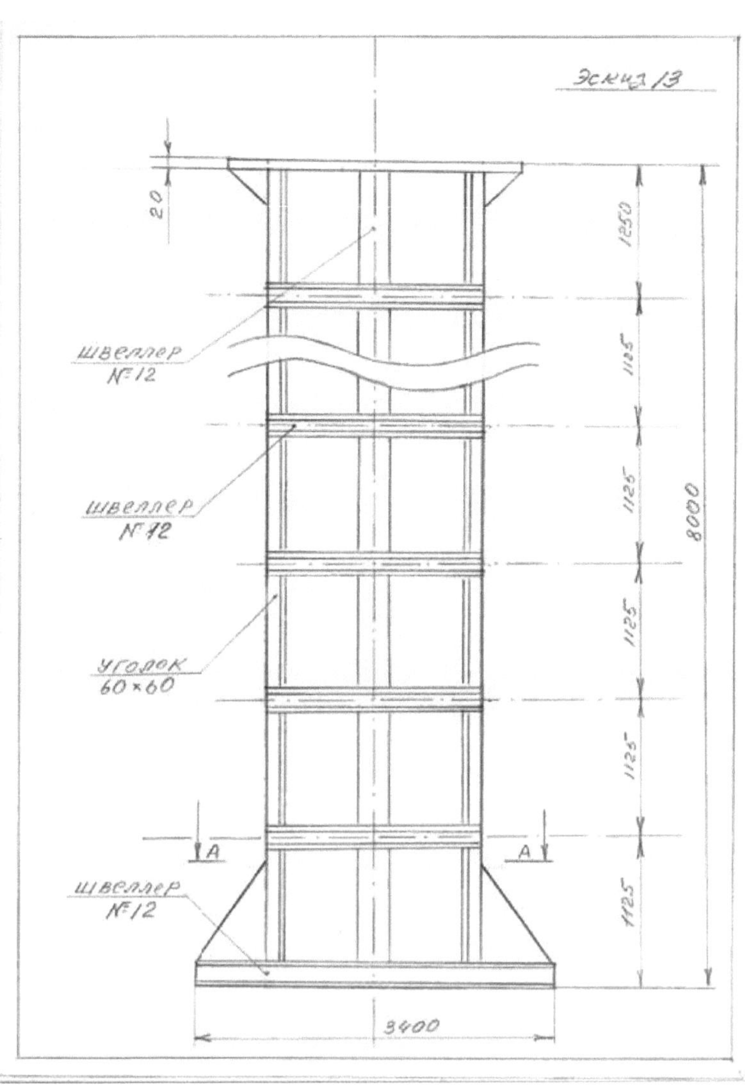
Эскиз 13
20
1250
1125
1125
1125
1125
1125
8000
швеллер
№12
швеллер
№12
уголок
60×60
А
А
швеллер
№12
3400

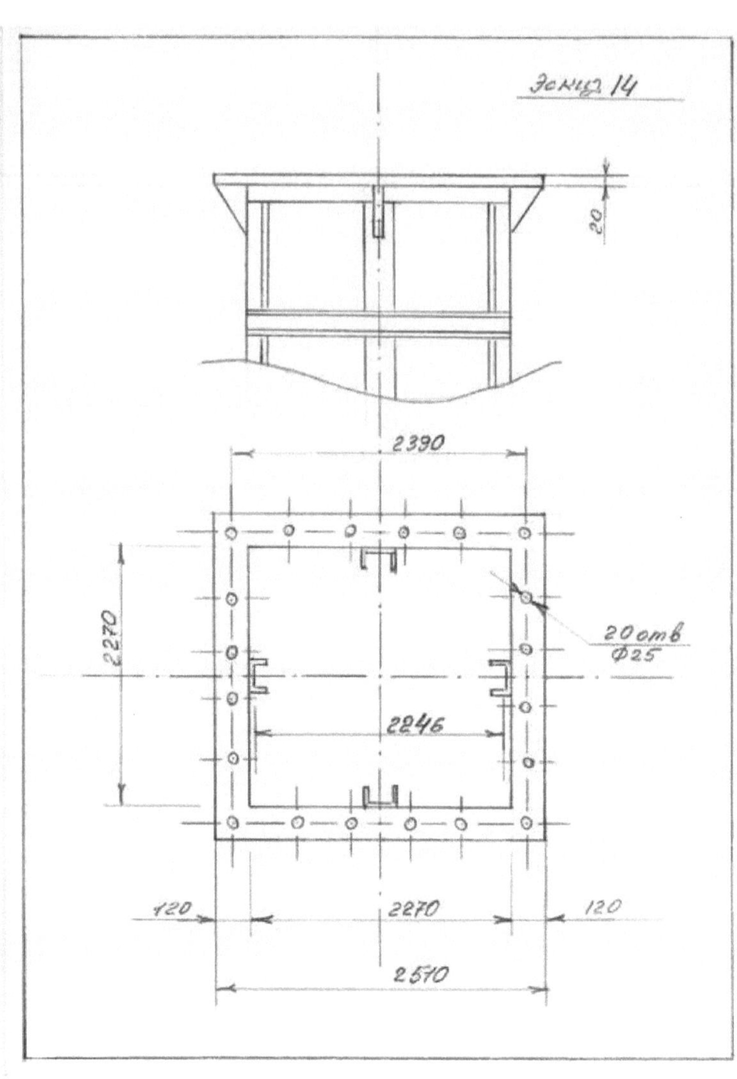
Эскиз 14
20
2390
2270
20 отв
Ф25
2246
120
2270
120
2510

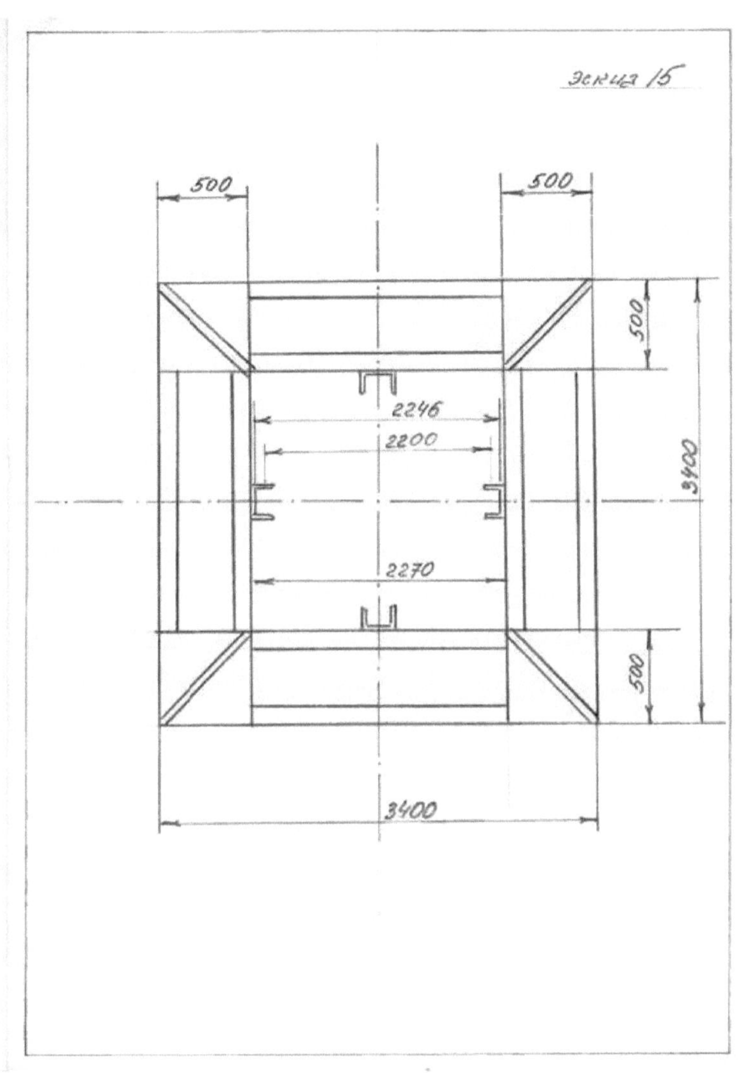
Эскиз 15
500
500
500
2246
2200
3400
2270
500
3400

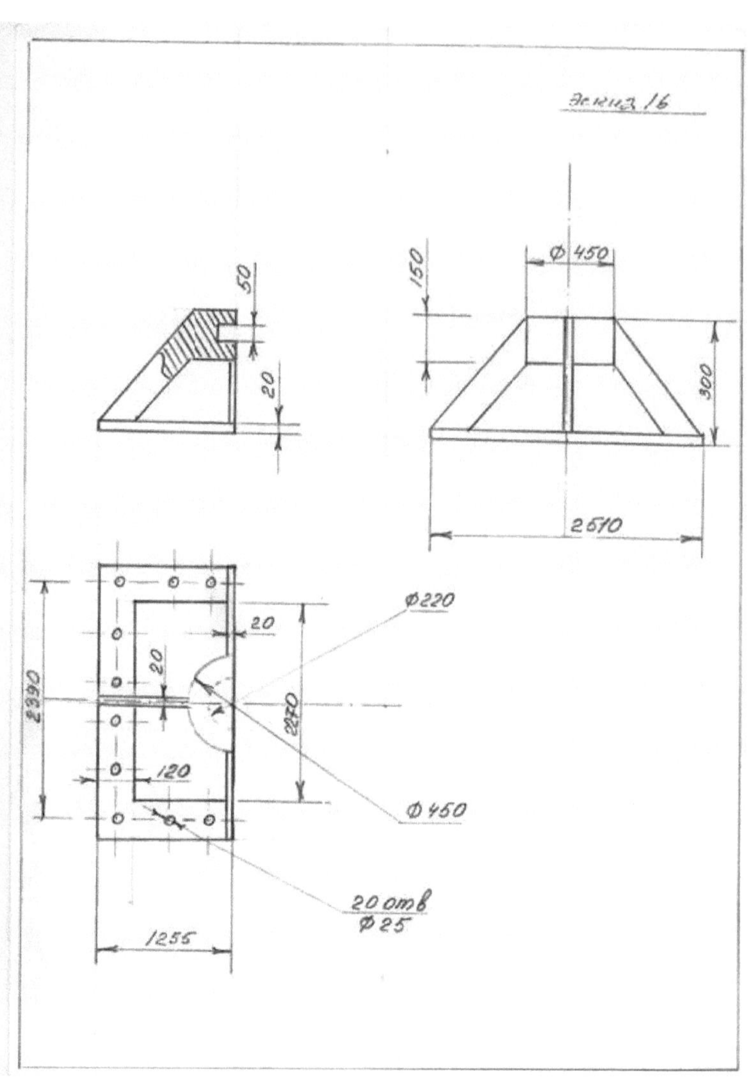
Эскиз 1б
50
20
150
Ф 450
300
2610
Ф220
20
20
2390
2270
120
Ф 450
20 отв
Ф 25
1255

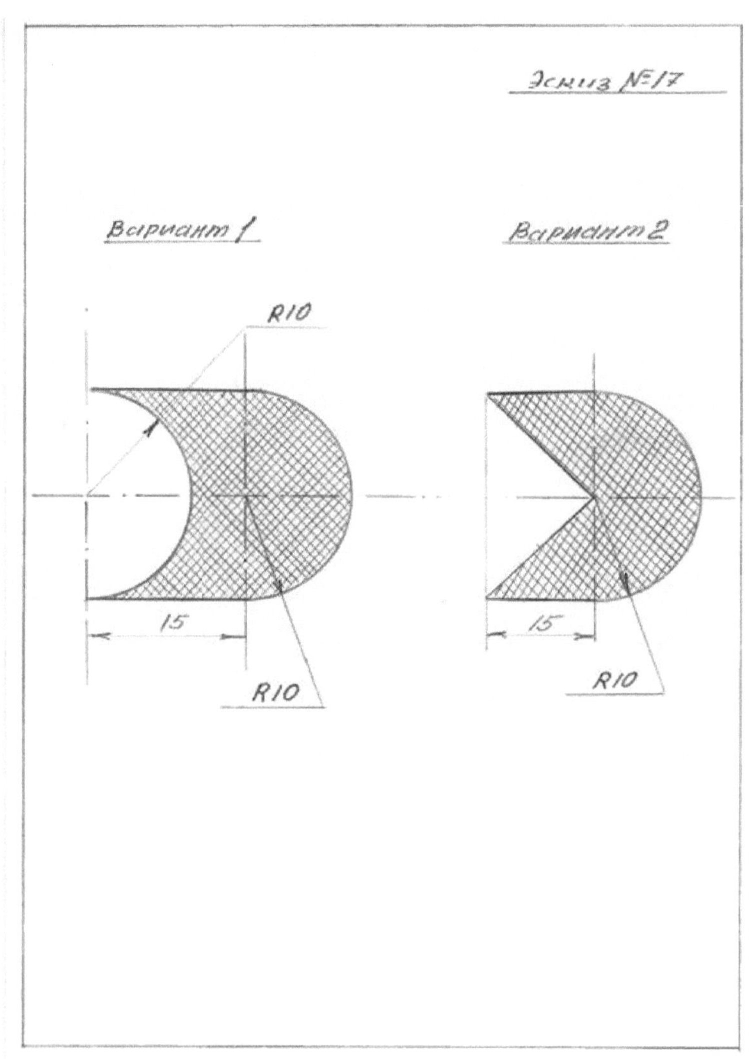
Эскиз №17
Вариант 1
Вариант 2
R10
R10
15
15
R10

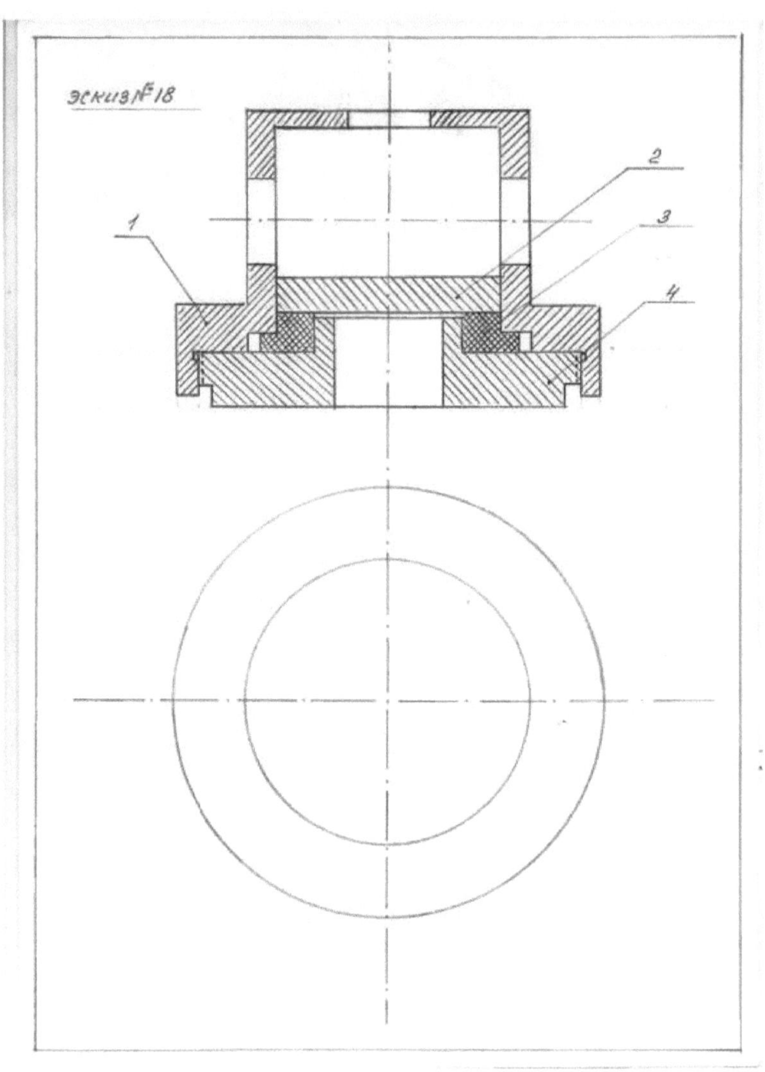
эскиз № 18
1
2
3
4

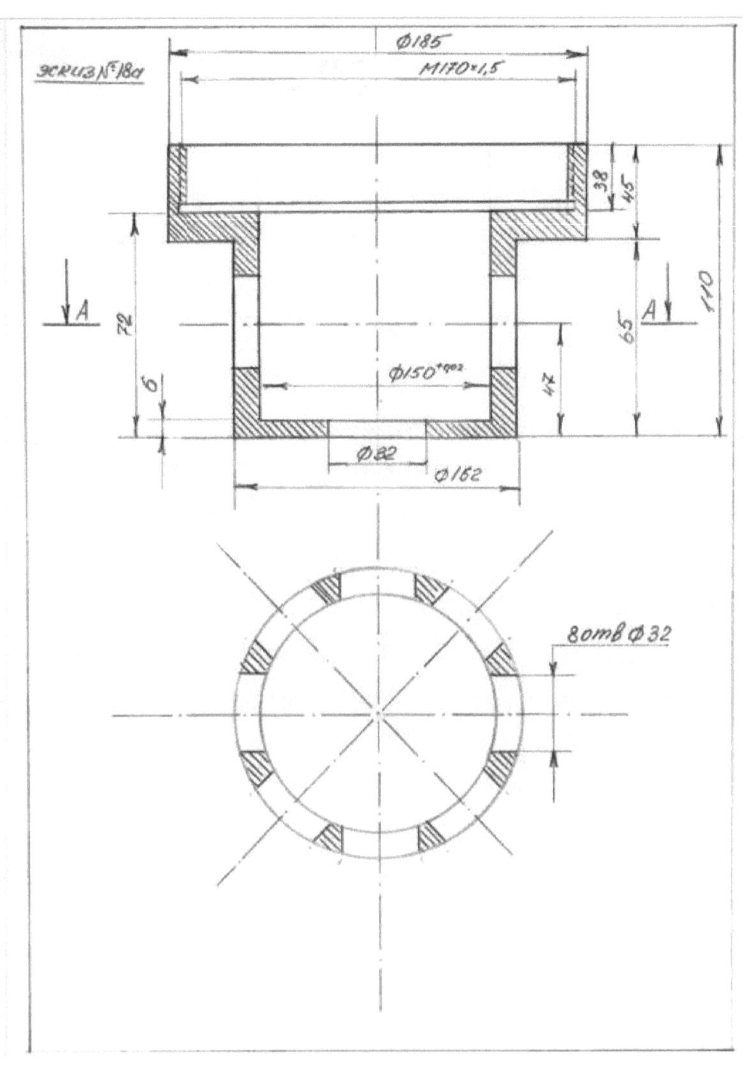
эскиз №18а
Ø185
M170×1,5
38
45
110
65
A
A
72
6
47
Ø150+0,02
Ø82
Ø162
8отв Ø32

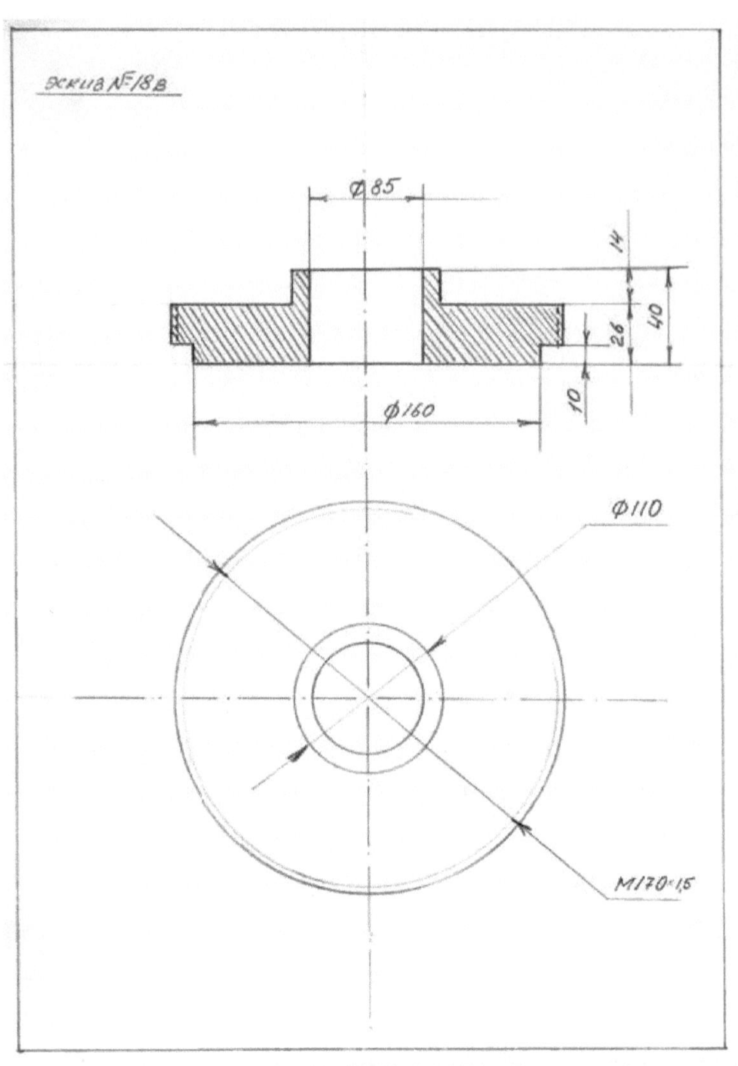
Эскиз №18в
Ø85
14
40
26
10
Ø160
Ø110
M170×1,5

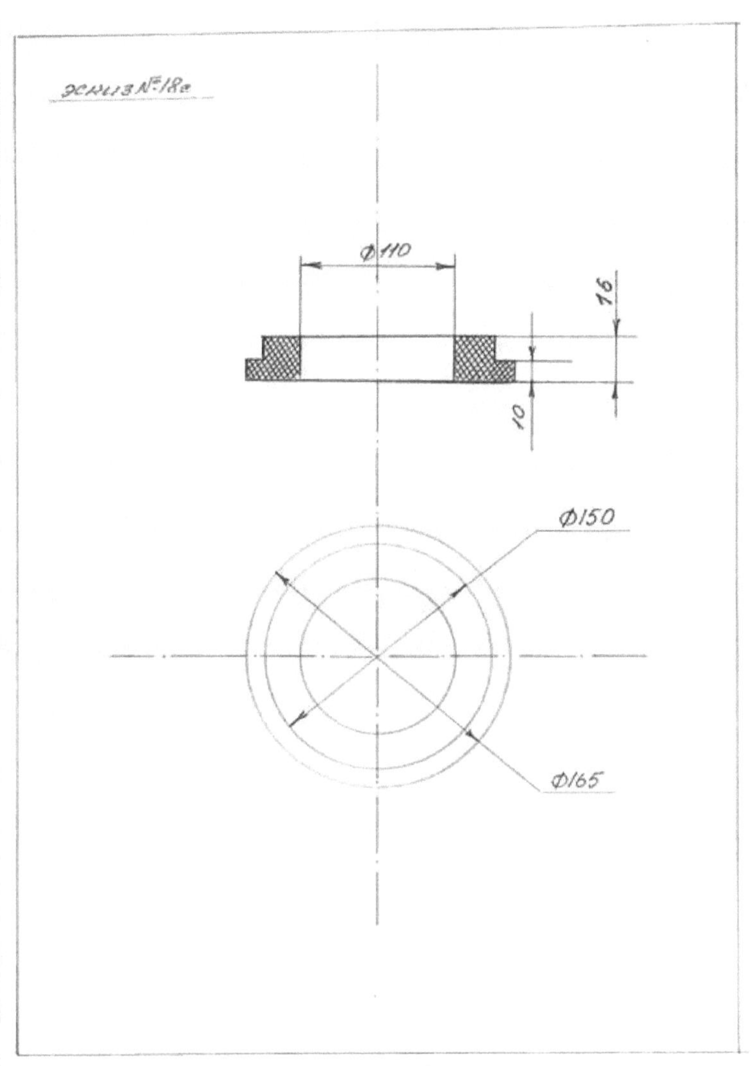
эскиз №18а
Ø110
16
10
Ø150
Ø165

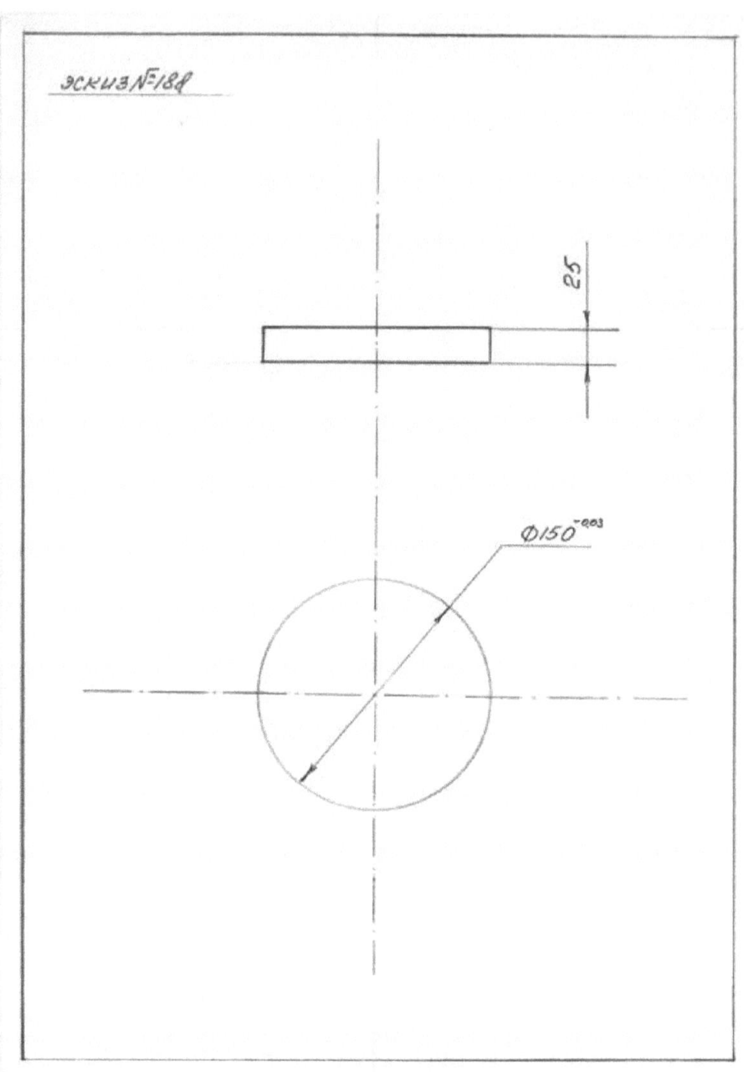
эскиз №18в
25
Ф150 -0,03

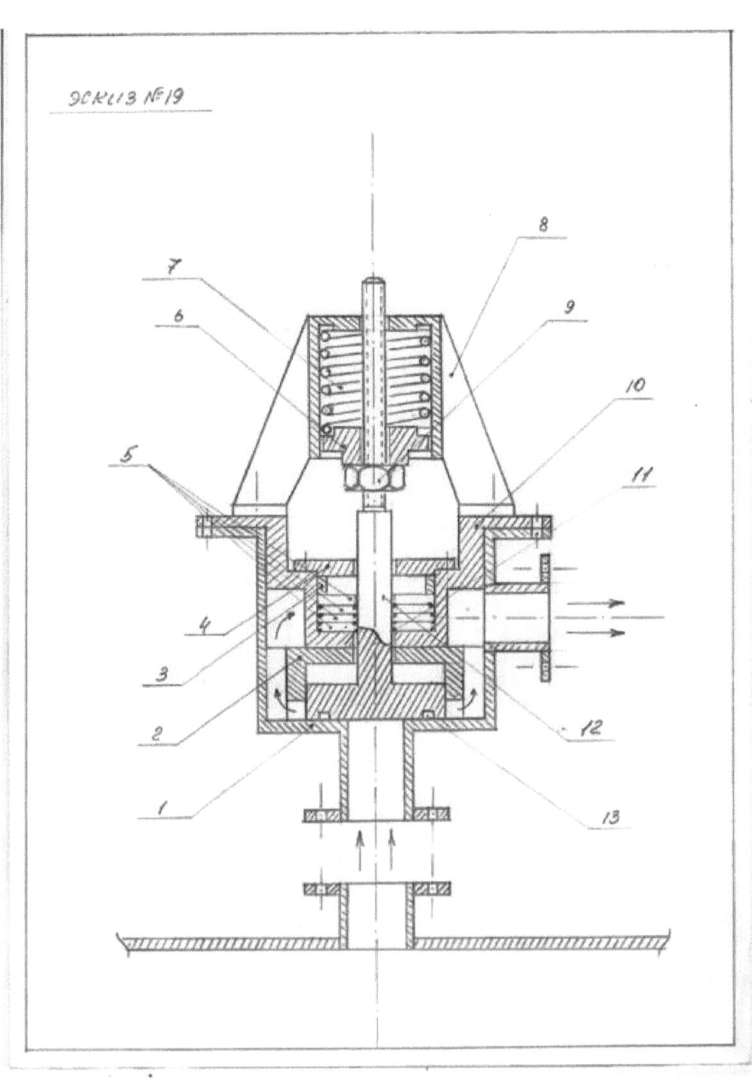
Эскиз №19
8
7
9
6
10
5
11
4
3
12
2
1
13

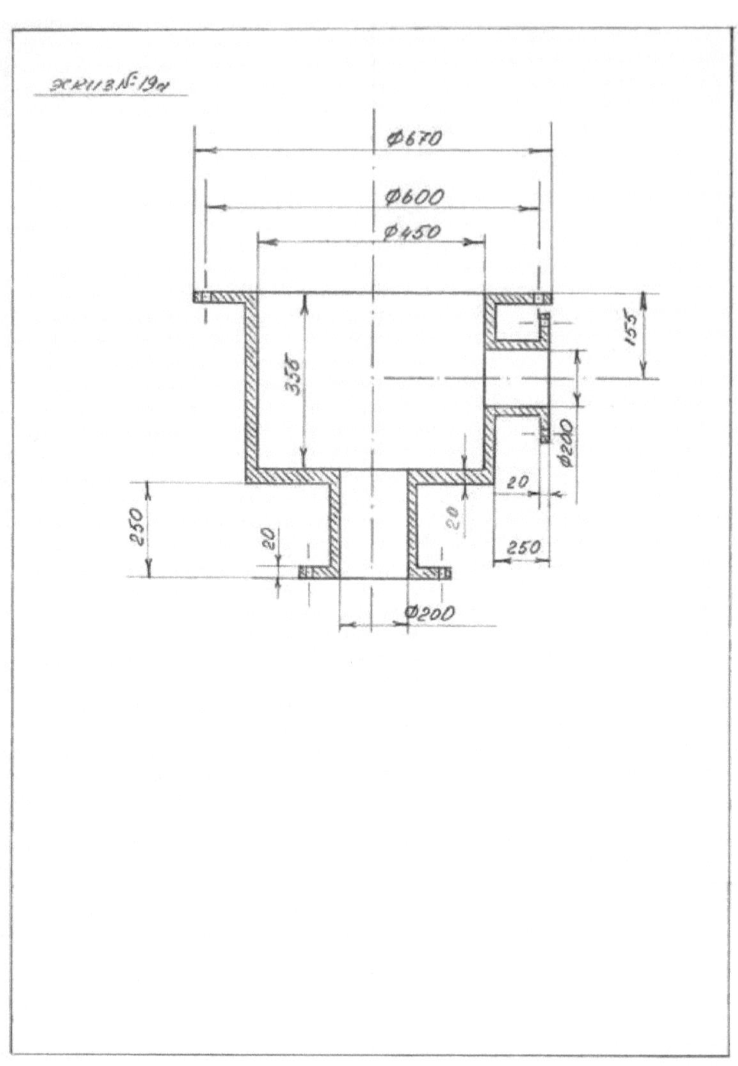
Эскиз № 19н
Ø670
Ø600
Ø450
155
350
Ø200
20
20
250
250
20
20
Ø200

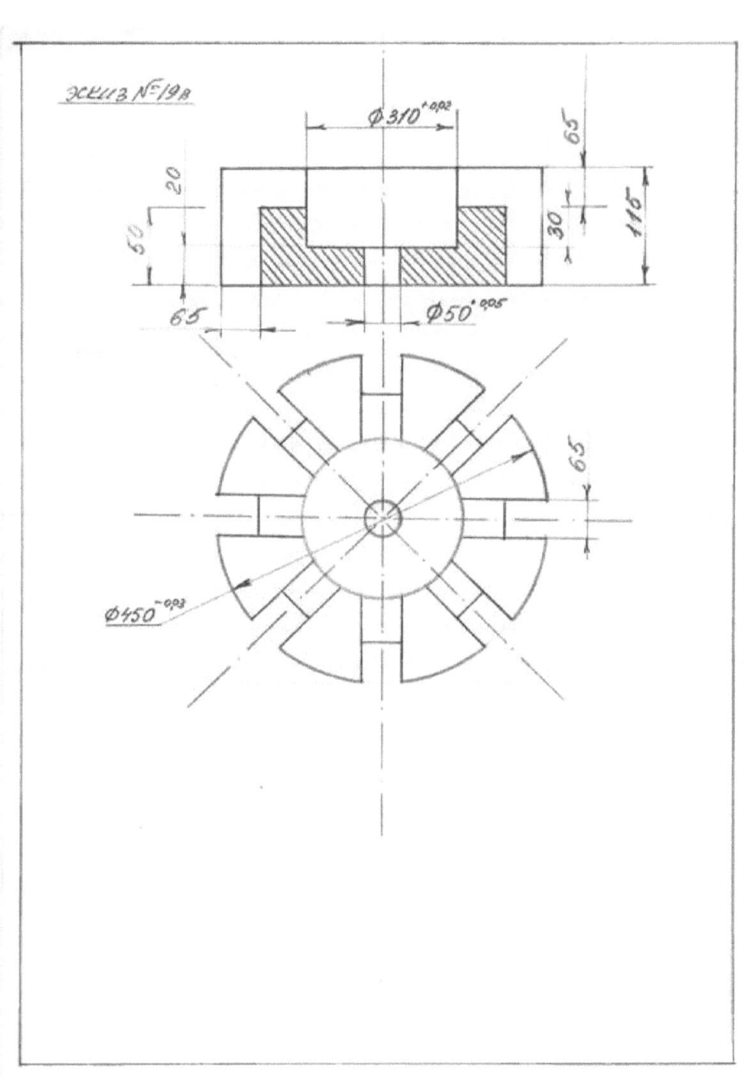
эскиз №19в
Ø310+0,02
65
20
30
115
50
65
Ø50+0,05
65
Ø450-0,03

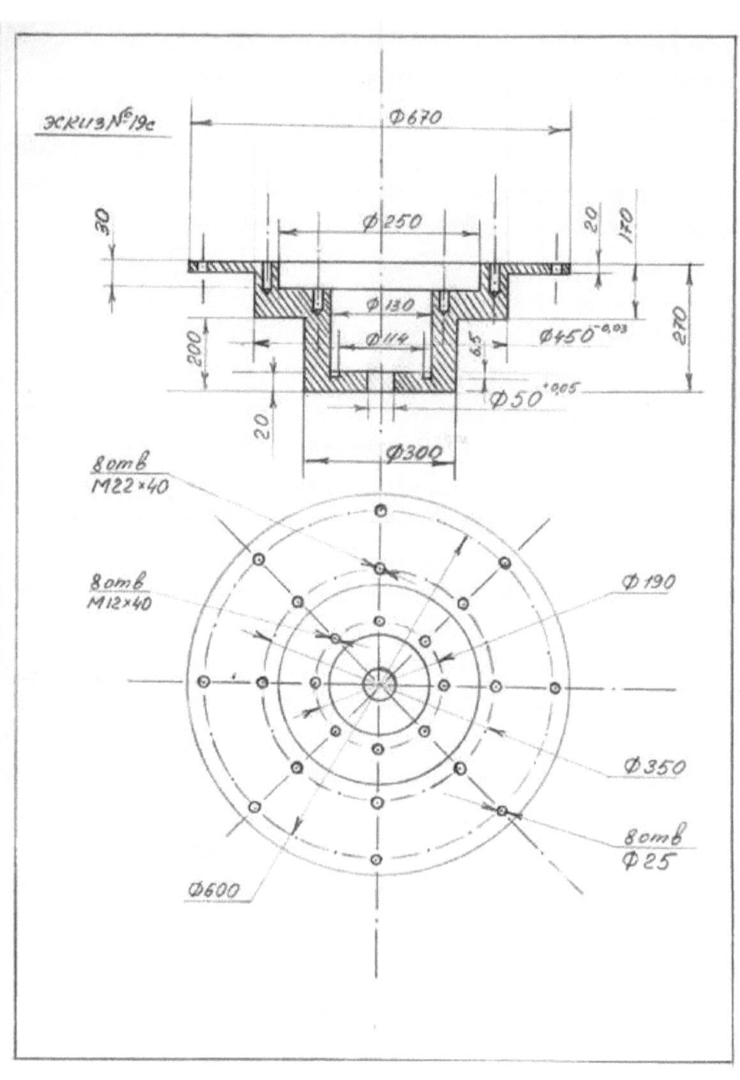
Эскиз №19с
Φ670
Φ250
Φ130
Φ114
Φ450$^{-0,03}$
Φ50$^{+0,05}$
Φ300
30
20
170
270
200
20
6,5
8 отв
М22×40
8 отв
М12×40
Φ190
Φ350
8 отв
Φ25
Φ600

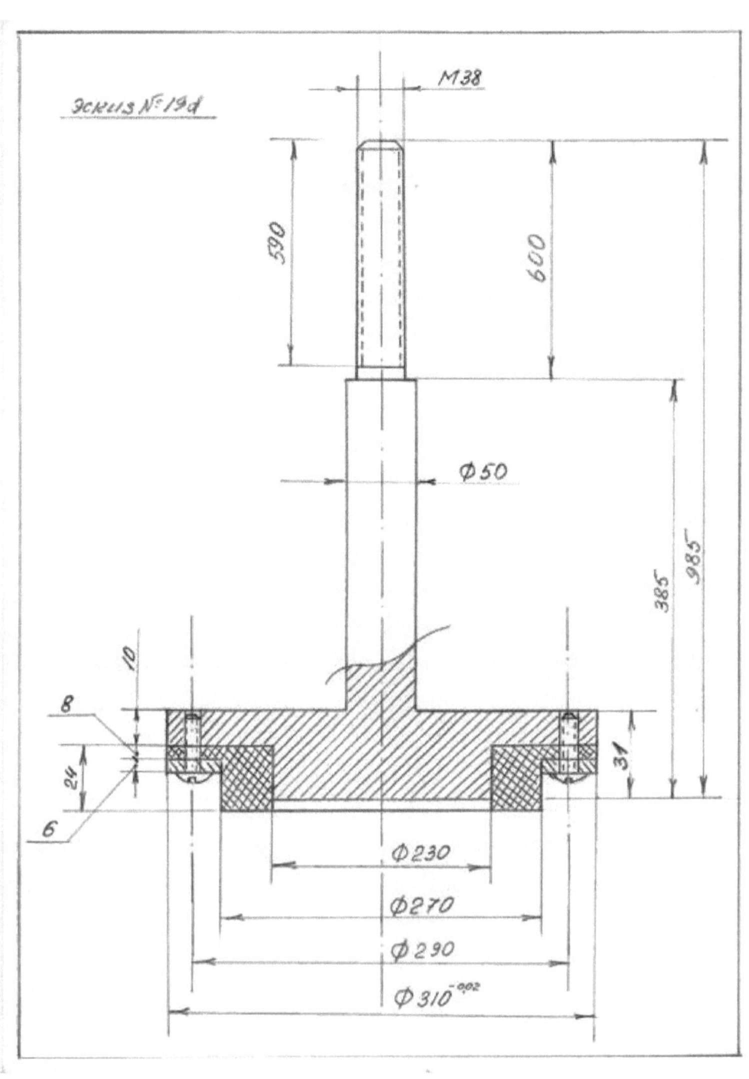

Эскиз № 19d
M38
590
600
Ø50
385
985
10
8
24
6
31
Ø230
Ø270
Ø290
Ø310

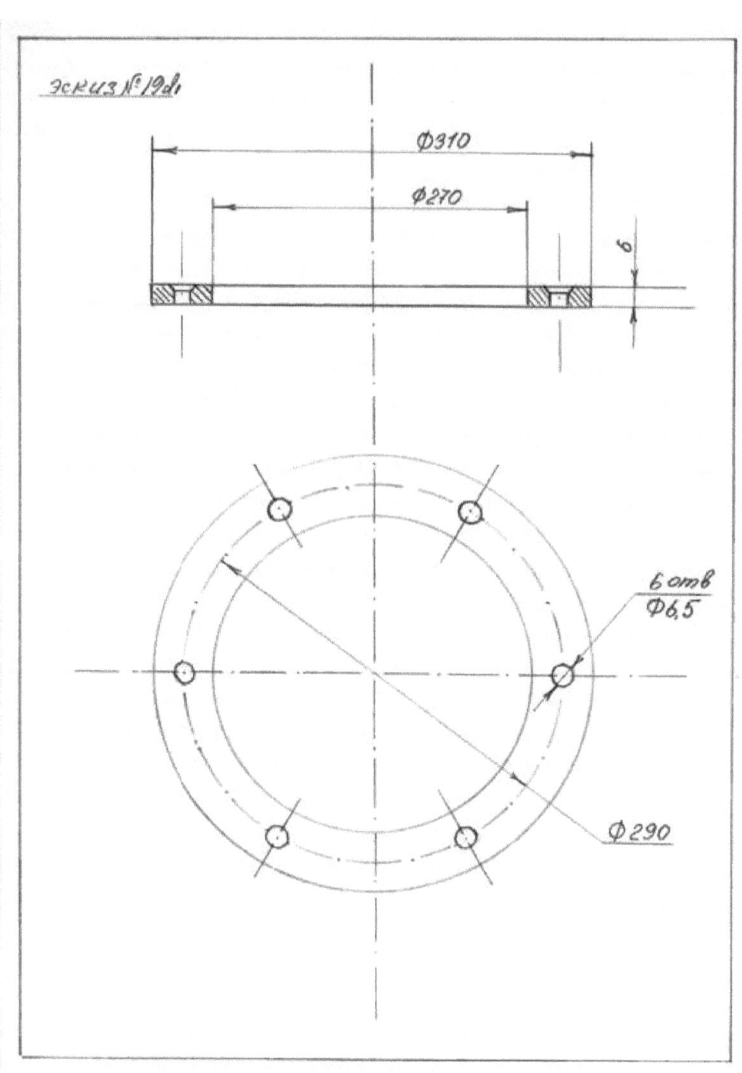
Эскиз №19д
Ф310
Ф270
6
6 отв
Ф6,5
Ф290

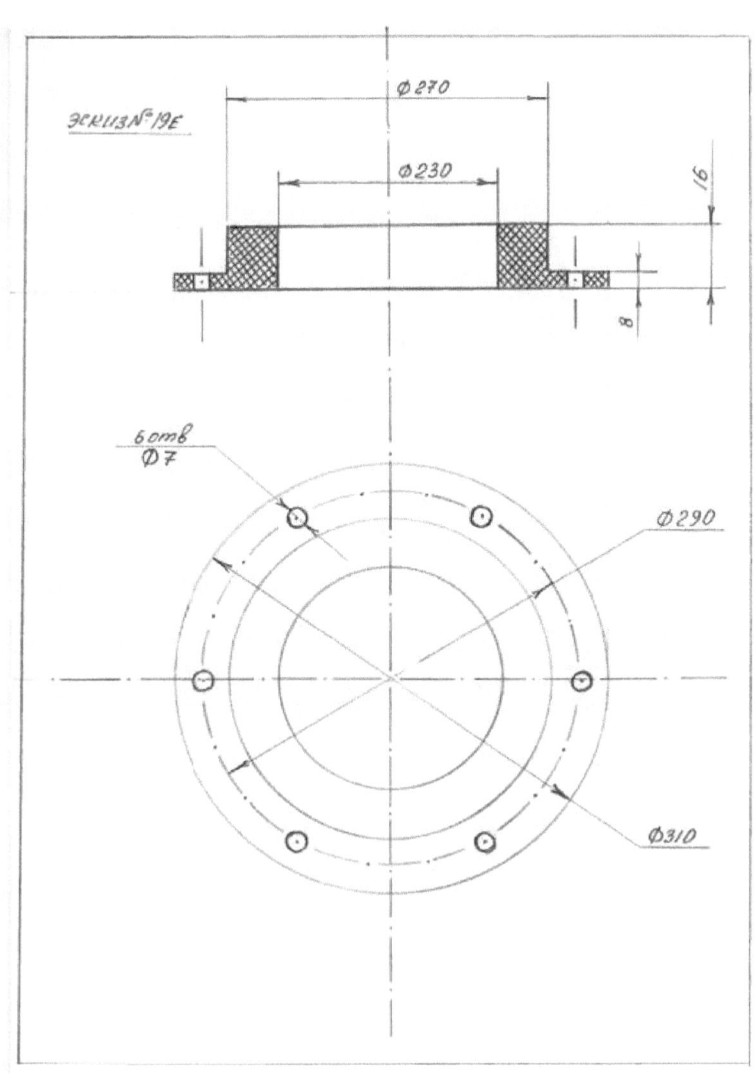
Эскиз № 19Е
Φ270
Φ230
16
8
6 отв
Φ7
Φ290
Φ310

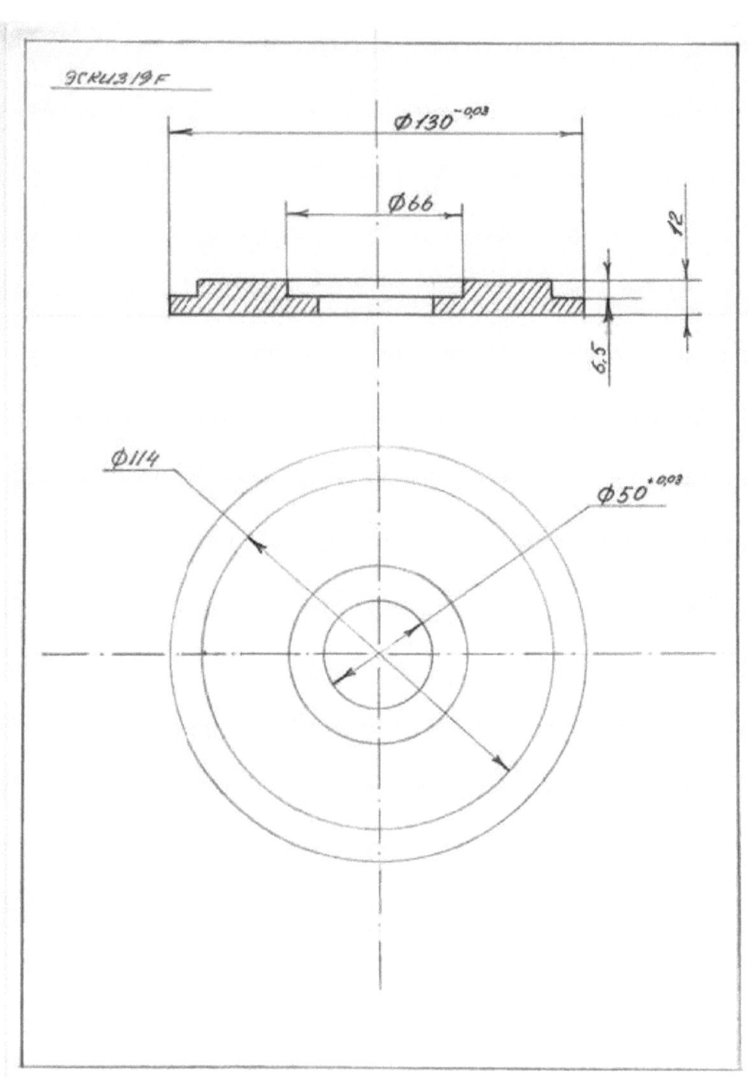
Эскиз 19F
Ø130$^{-0,03}$
Ø66
12
6,5
Ø114
Ø50$^{+0,03}$

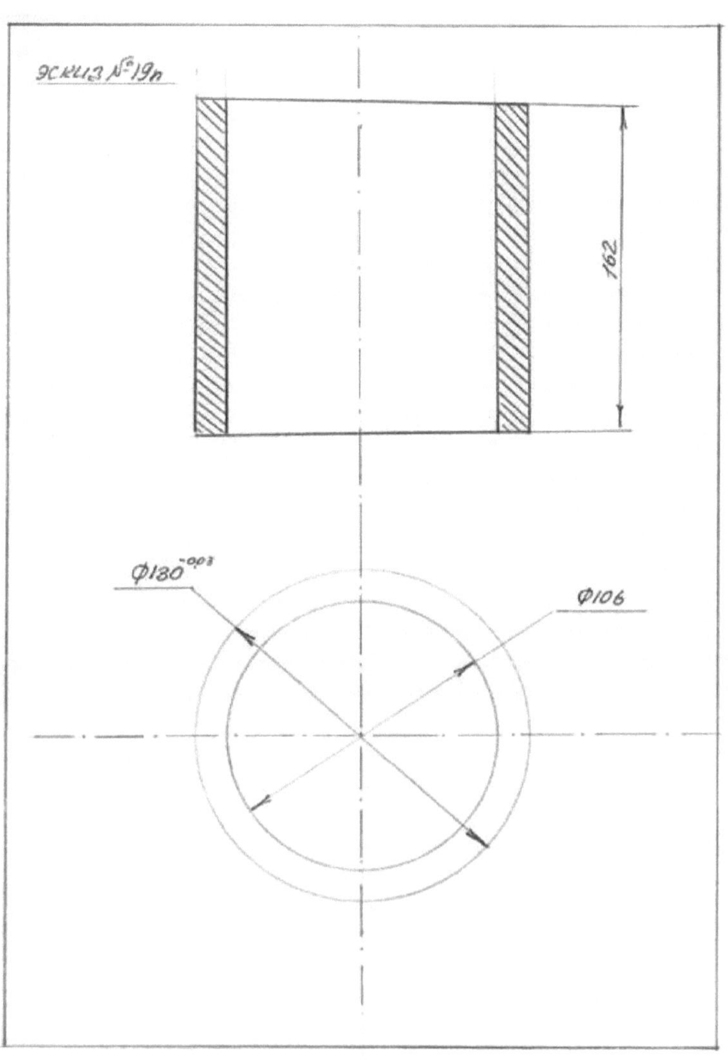
эскиз №19п
162
Ф180-0,03
Ф106

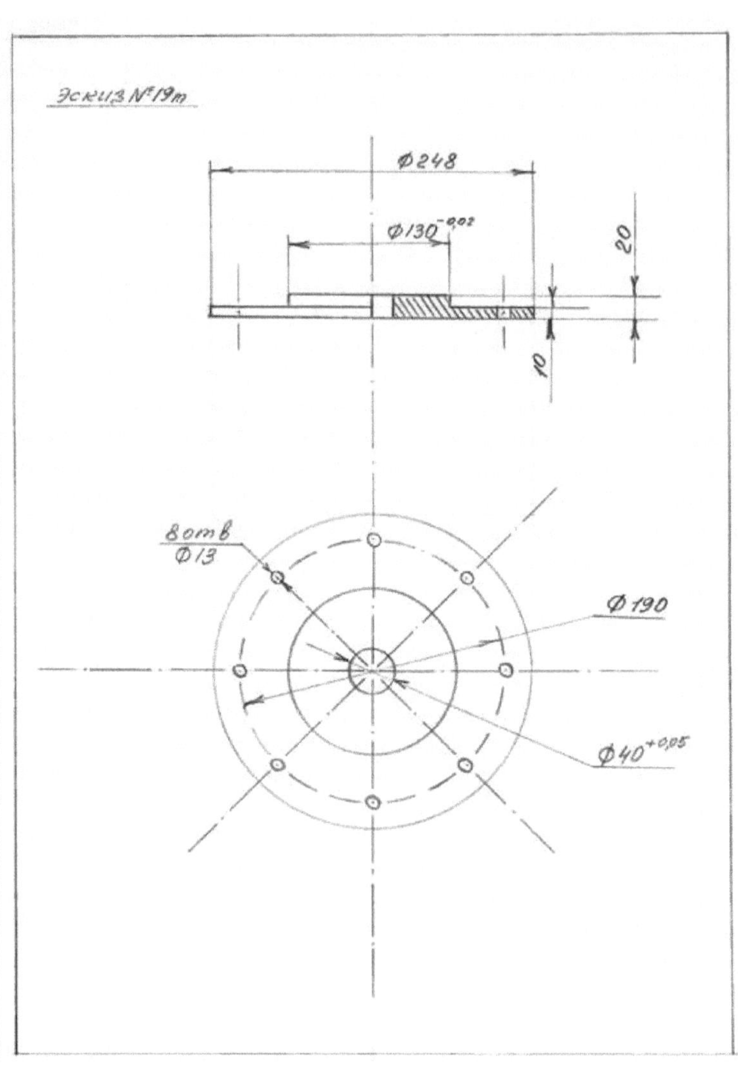
Эскиз №19т
Ø248
Ø130-0,02
20
10
8отв
Ø13
Ø190
Ø40+0,05

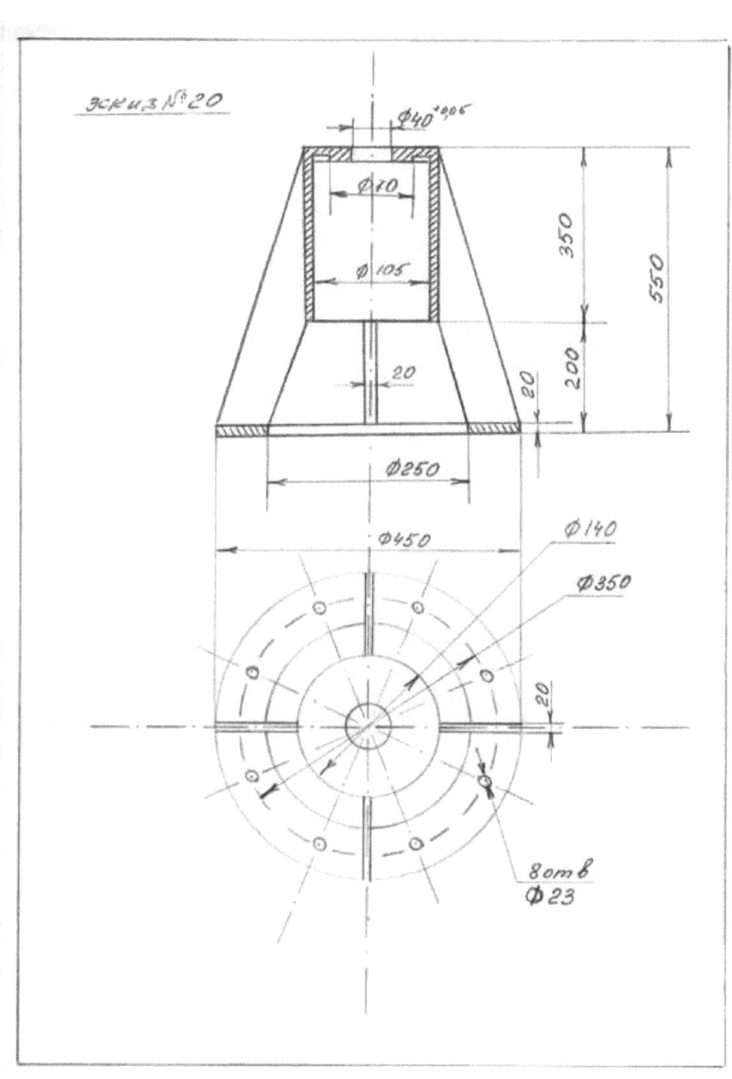
эскиз №20
Φ40+0,05
Φ70
Φ105
350
550
200
20
20
Φ250
Φ450
Φ140
Φ350
20
8 отв
Φ23

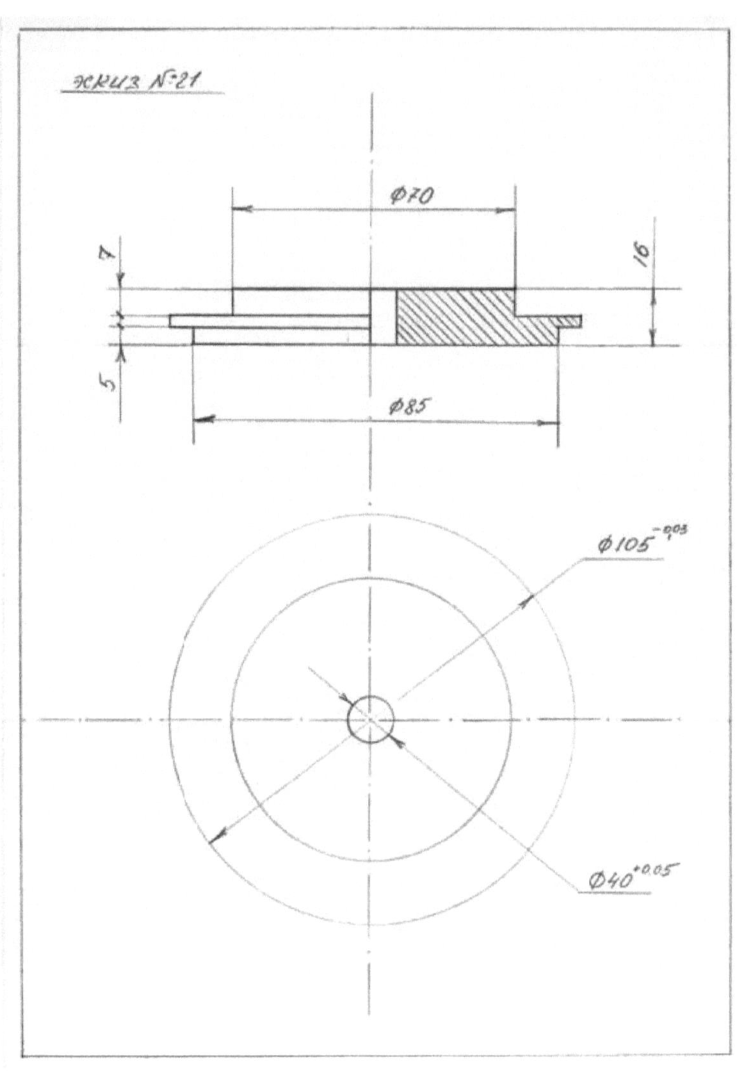
Эскиз №21
Ø70
7
16
5
Ø85
Ø105⁻⁰·⁰³
Ø40⁺⁰·⁰⁵

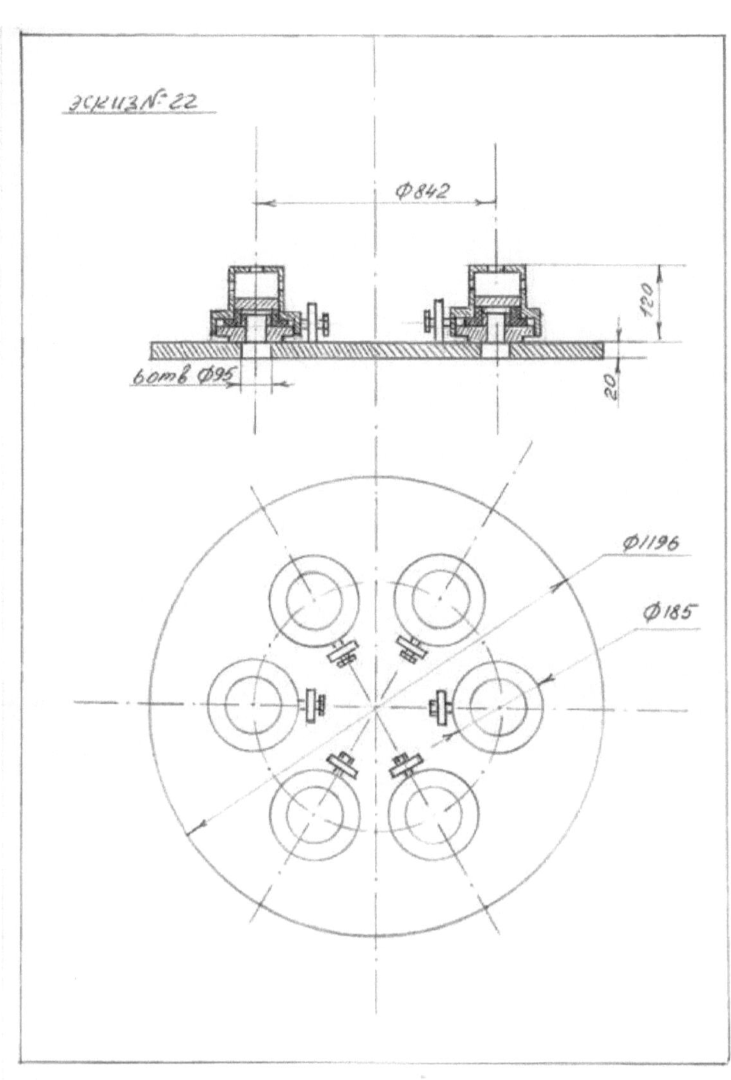
Эскиз № 22
Ф842
120
6отв Ф95
20
Ф1196
Ф185

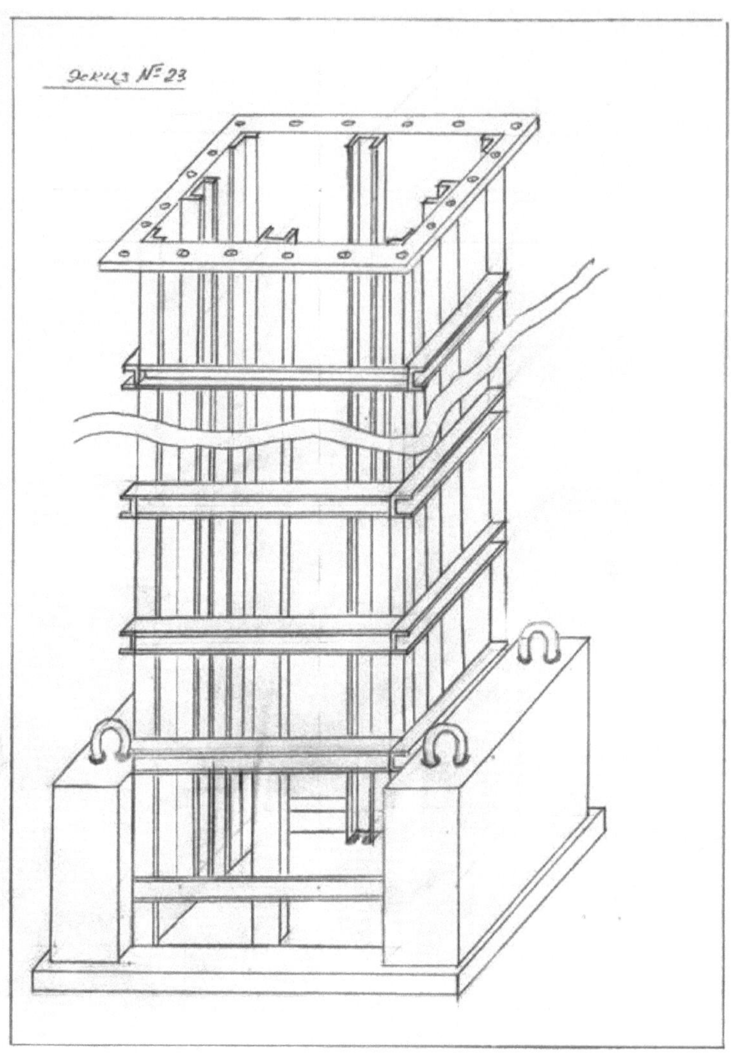

According to Table 1 Table 2, the total amount required for materials for the pump section is 70671.95man. Or 39585.5$

CHAPTER 2

HYDROPOWER AND GRAVITATIONAL ENERGY, CAN SOLVE THE ENERGY PROBLEM.

Azerbaijan Research Institute of Geotechnical Problems of Oil, Gas and Chemistry Author: Engineer of Alternative Energy Laboratory
Boris Vladimirovich Silvestrov boris silvestrov@mail.silvestrov@mail.ru

Hydropower is an environmentally friendly type of electricity generation. Today, the potential for hydropower generation has been practically exhausted. It has become impractical to build dams and flood vast fertile territories. Besides, damming rivers causes ecological damage to fauna. The spawning of valuable fish species is hindered. Navigation on the rivers is experiencing difficulties. It is necessary to build expensive locks. And yet there is an opportunity to give hydropower a second breath by completely abandoning the construction of dams. Quite powerful hydroelectric power plants can be built wherever there is a body of water. It can be a river, canal, lake, sea, pond, ocean. In other words, almost everywhere on planet Earth where there is water and it does not matter whether it is salty or fresh. It is this method of electricity generation can give mankind the necessary clean energy in unlimited quantities. This is what will be discussed in this article. The main pioneering equipment to implement this method is a two-way pontoon pump with a fixed piston. This pump is capable of converting gravitational energy (Weight) into water jet energy and further into electricity. With its help it is not only possible to obtain a rather large capacity, but also to achieve a very high pressure of the water jet fed to the blades of the hydro turbine, and this makes it possible to use powerful, high-pressure hydro turbines "Pelton" for the production of electricity. The water from the reservoir for the operation of the two-way pontoon pump will be supplied by a centrifugal pump with the appropriate, necessary capacity. The power spent on the work of the centrifugal pump is many times less than the power that can be obtained from the work of the pontoon two-way pump. This assertion will be proved in this article by calculations, according to the parameters of a particular sample of a pontoon pump, and taken for comparison from the catalogue, the parameters of a centrifugal pump. Sketch No. 1 shows a two-way, pontoon pump

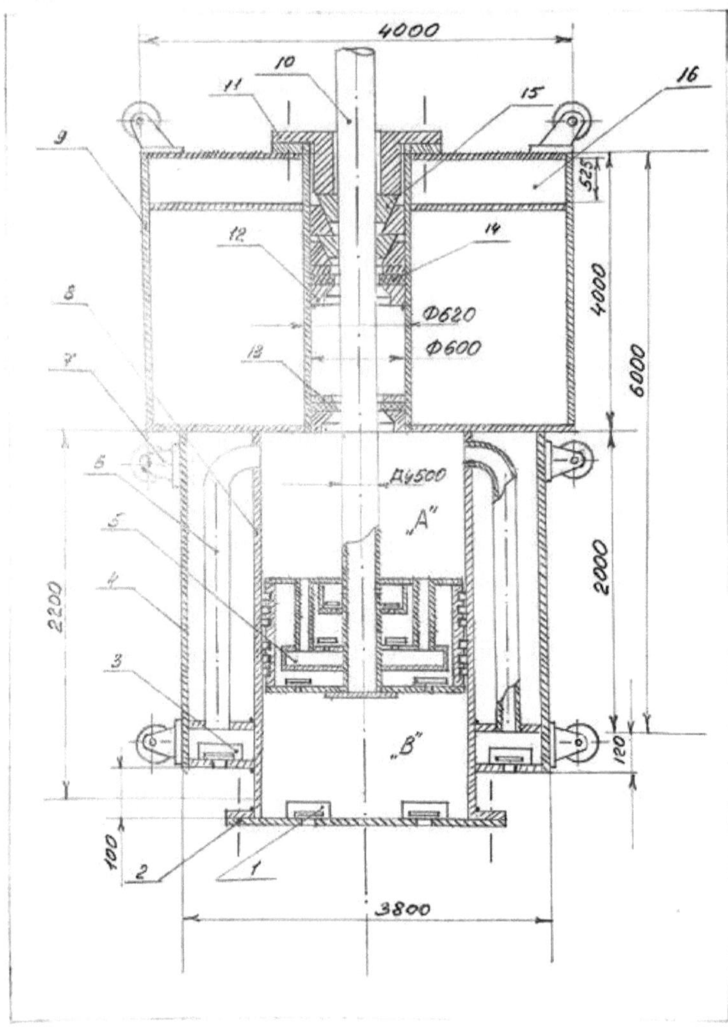

Sketch No. 1
Where:
1) Outboard water inlet valves of chamber "B"
2) Lower slave cylinder cover
3) Outboard water inlet valves of chamber "A"
4) Lower part, housing of the double-way pontoon pump
5)Pump fixed piston head
6) Water pipes to chamber "A"
7)Guide rollers
8)Working cylinder of the double-way, pontoon pump.

9)The upper, floating part of the pontoon.
10)The water line of the fixed piston.
11)Push-pull grandbux.
12) Support ring.
13)Retaining ring.
14) Copper guide ring.
15)Cone seals, caprolon seals.
16) Ballast water tank.

A two-way fixed piston pontoon pump operates by cycling the water level. It delivers a high-pressure jet of water to the blades of the Pelton bucket hydroturbine, which contributes to power generation.

Let us consider the operation of the piston pump in more detail. Sketch No.1 shows that when the cylinder, together with the pontoon, starts to rise together with the water level increase, a certain vacuum is created in chamber "A". At the same time, the overboard water valves (3) open and water fills chamber "A". In contrast, chamber "B" is pressurised. The outboard water valves (1) close and the pressurised water flows through valve (A) and then through valve (C), (see Fig. 3) into the conduit and further into the compensation column, while valve (B) (Fig. 3) closes, preventing water from entering chamber "A". Further, when the pontoon together with the change of water level starts to sink, then, on the contrary, in the chamber "B", a certain vacuum will be created, the valves of the sea water will open, filling the chamber "B" with water. In chamber "A", this will create pressure and water through valve (B). And then through the valve (C), (Fig. 3) will again get into the water line, at the same time the valve (A) (Fig. 3) will close and prevent the flow of water into the chamber "B". The valve "C" will also close, preventing water outflow from the conduit, until the moment of squeezing and water exit through the valve "C" into the conduit.

The piston (Fig. No. 3) is a welded construction,

welded from elements made of 20mm and 100mm thick sheet iron and 200mm diameter pipes.

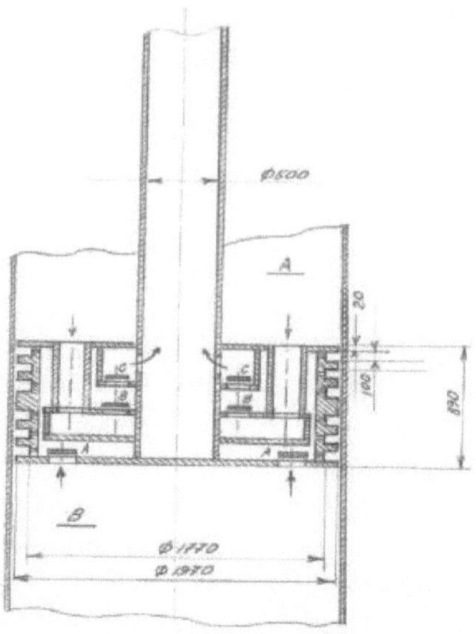

Fig. 3

To evaluate the performance parameters of the proposed pontoon pump design, it is necessary to calculate its capacity and the maximum pressure it can generate. Knowing the flow rate and pressure it is possible to calculate the capacity of the proposed high-pressure hydraulic unit by the formula: W= flow rate (m^3 /sec) x head (kg/cm^2) x 9.8 m/sec^2 x 0.6

To consolidate the above material, let us consider the operation of one proposed module with the dimensions of the pontoon pump casing (shown in sketch #1) equal to 3.8m x 3.8m x 2m. plus the upper, additional, float tank 4m x 4m x 4m x 4m. The diameter of the piston is 2m (see sketch #1). The water level fluctuation in the buried tank is 1.2m. (see sketch #2)

Let's make calculations according to the above mentioned parameters: The total volume of the pontoon will be 85.174m^3 . After manufacturing the pontoon pump casing it is necessary to determine its weight. Then find the total weight of the moving pump casing and water in the working chambers. If this turns out to be less than half the volume, then add water to the special chamber located in the upper part of the pontoon so that the total weight of the material from which the casing is made and the weight of water in the compression chambers and the weight of water in the special chamber is equal to half of the air volume.

Adopted parameters for the calculation of the power plant:

Piston diameter 2м.

Cylinder stroke calculation 1.2м

Water level fluctuation period 6sec.

Radius of the water pipe 0,5м

SECTION CAPACITY CALCULATION

The capacity of the chamber (A) is equal to the volume of water displaced by the piston into the conduit, at the accepted displacement of the cylinder, which in turn is equal to 1.2m

$$Va = \pi r_1^2 h - \pi r_2^2 h$$

$$Va = 3{,}14 * 1м^2 * 1.2м - 3{,}14 * 0{,}25^2 * 1.2м = 3{,}768м^3 - 0{,}2355м^3 = 3.533м^3$$

Where: r1 - radius of the slave cylinder

h, - height of cylinder displacement

r2, - radius of the water pipe.

The capacity of the chamber (B) is respectively equal to the volume of water displaced into the conduit at a cylinder stroke of 1.2m.

$$V_B = \pi r^2 h$$

$$V_B = 3{,}14 * 1м^2 * 1.2м = 3{,}768м^3$$

The total volume of water in one cycle (moving the cylinder up and down 6sec) will be:

$$V_\Sigma = Va + V_B = 3.533м^3 + 3{,}768м^3 = 7.3м^3$$

Taking into account that the periodicity is equal to 6sec. Let's find the productivity of the section for one second.

Thus, the capacity of the section in one second when moving the cylinder 1.2m is equal to:

7.3 m3 : 6sec = 1.21m3/sec

Thus, the total water flow through the buried tank can be assumed to be from $1.5m^3$ /sec to $2.5m^3$ /sec. This is quite sufficient for the operation of one hydropower module. The smaller the gap between the casing of the pontoon pump and the chamber in which it operates, the smaller the difference between the water supplied by the centrifugal pump and the water required for the operation of the pontoon pump.

Next we calculate the air volume of the pontoon part of the pump.

The volume of the lower section is equal to: the volume of this section minus the volume of the working cylinder, and the volume of the six outboard water connections to chamber "A"

$Q_H = 3.8m * 3.8m * 2m - 3.14 * 1m^2 * 2m - 6 x 3.14 x 0.10952m =^2$

$=28.88, m^3 - 6.28m^3 - 0.2258m^3 = 22.374m^3$

The volume of the upper part of the pontoon section will be equal to: the volume of the upper part minus the volume of the cylinder inside which the water pipe runs. (diameter 620mm)

$Q_B = 4m * 4m * 4m - 3.14 * 0.3W * 4m = 64- 1.2 = 62.8m3$

The total, air volume, will be equal to:

$Q_\Sigma = Q_H + Q_B = 22.374м^3 + 62.8м^3 = 85.174м^3$

The constant volume of water inside the cylinder when the pump is running is equal to the working volume of the cylinder minus the volume of the piston head:

$V_B = \pi r^2 h = 3.14 \times 1^2 м^2 \times 2м - 3.14 \times 1^2 м^2 \times 0.89м = 6.28м^3 - 2.795м^3 = 3.485m^3$

or

that equals 3,485 tonnes of water.

The combined weight of the moving part of the pump casing and the water in the working cylinder as well as the water in the ballast chamber should be equal to half of the air volume, which in turn ensures the buoyancy of the pump casing.

It is necessary to design the moving part of the pump so that its total weight and the weight of the water in the working cylinder and in the ballast chamber are equal to half of the volume, then the residual buoyancy will also be equal to half of the volume. The pump will run with equal force in both directions. In Pelton high-pressure hydraulic turbines, at maximum dimensions, the maximum number of nozzles is 6, and the maximum diameter of each nozzle, in the open state, is 80mm. Then the total area of the expelling water jet is equal to:

$S = \pi r^2 \times 6 = 3{,}14 \times 4^2 см \times 6 = 301{,}44 см^2$

Calculation of the weight of the moving part of the pontoon pump

1)Weight of cylinder (piston diameter 2m cylinder height 2.3m wall thickness15mm =0.015m)

The length of the circle is $2\pi r = 2 \times 3{,}14 \times 1 = 6{,}28м$

The volume of the cylinder is $V = 6.28m \times 2.3m \times 0.015m = 0.217m^3$

The weight of the cylinder is $P = V \times 7.8t/m^3 = 0.217m^3 \times 7.8t/m^3 = 1.692t$

2) The weight of water pipes in the chamber "B" 8 pipes diameter 219mm (Theoretical weight of 1 linear metre of steel electrically welded pipe 0 219h8 - **41.74 kg.**) The length of one pipe 2.05m. Total length of pipes 16,4m weight is P= 16,4 x 41,74 = 684,5 kg = 0.685t

3)Weight of four side walls of the bottom box (each wall 2.0m x3.8m total area $S=30.4М^2$ with a sheet thickness of 15mm the volume is $0.456m^3$ Weight=3.557t

4)Weight of the upper partition at the water inlet to chamber "B"

Area $S = 3.8m \times 3.8m - pg^2 = 14.44m^2 - 3.14 \times 1^2 m = 11.3m^2 - (8 \times 3.14 \times 0.1095^2 m) = 11m^2$ with a partition thickness of 15mm Weight will be equal to $P = 11m^2 \times 0.015m \times 7.8 t/m^3 = 1.287t$

5)Weight of the lower partition at the water inlet to chamber "B"

As with the top bulkhead, the weight will be 1.287tn

6)Weight of the lower cylinder head

With a cover thickness of 15mm, the diameter is 2.2m minus the holes for the inlet valves (219mm 8 pcs).

S= 3.14 x 1.12m - (8 x 3.14 x 0.10952m) = 3.799m^2 - 0.301m^2 =3.498m^2

Volume V = 3.498m^2 x 0.015m = 0.052m^3

Cover weight P = 0.052m^3 x7.8t/m^3 =0.409 tonnes

7)Weight of bottom flange outer diameter 2.2m inner diameter 2.0m flange thickness 50mm (3.14 x 1.1^2 m) - (3.14 x 1^2 m) x 0.05m x 7.8t/m^3 =(3.799m^2 - 3.14m^2) x 0.05m x 7.8t/m^3 = 0.257t

8) Weight of inlet valves

Total 16 pcs. valves weight each of the order of 4 kg total weight of valves is equal= 4 x 16 = 0,064t

9) Weight of rollers total 24 rollers each weighing approximately 6kg

Total weight of rollers P= 24 x 6 =144kg = 0, 144t

10)Weight of vertical walls of upper float chamber 4m x 4m x 4 = 64m^2 with wall thickness of 15mm weight of vertical walls is P = 64m^2 x 0,015m x 7,8t/m^3 =7,488t

11) Weight of inner pipe pipe diameter 0.62m wall thickness 10mm (0.01m) height 4m. Total surface area = S=2 x 3.14 x 0.31m x 4m=7.787m^2 volume is V=7.787m2 x 0.01m =0.0778m^3 then weight= 0.0778m^3 x 7.8t/m^3 = 0.607t

12) Weight of the upper lower and middle plane on the float part of the pontoon

Each plane is a 4m x4m square minus the area of a circle with a diameter of 620mm with a thickness of 15mm top and bottom middle 10mm

S = 4m x 4m - 3.14 x 0.312m=15.698m2 V = 15,698m^2 x 0,015m =0,235m^3

+Weight of two planes P = 2 x 0.235m^3 x 7.8t/m^3 = 3.673t

partition volume V= 15.698m^2 x 0.01m =0.156m^3

partition weight P = 0.156m^3 x 7.8t/m^3 = 1.224t

13) The weight of the hailbuckle and seals is approximately 0.25t

Let's calculate the total weight of the float pump moving part, it will be summarised by the following weights:

1)Weight of slave cylinder1, 692t

2) Weight of water pipes to chamber "B" 0.685t

3)Weight of four walls of the bottom box3, 557t

4)Weight of the upper partition at the entrance to chamber "B" 1,287t

5)Weight of the lower partition at the water inlet to chamber "B"

1,287т

6)Weight of the lower cylinder head0 .409t

7)Weight of inlet valves0.064t

8)Weight of rollers0 .144t

9) Weight of vertical walls7 ,488t
10)Weight of inner pipe0 .607t
11)Weight of upper and lower plates 3.673t
12)Weight of partition wall1 ,224t
13) Weight of grand axle and seals0 ,250t

Total: 22,624t

According to the above calculations, the weight of water in the working chambers of the cylinder is 3.485t. Thus the total weight of water in the chambers and the weight of the movable part of the pontoon will be equal to P=22.624t +3.485t = 26.109t. According to the fact that the weight of the movable part should correspond to half of the volume, it is necessary to add some water to the ballast tank. Let's calculate the amount of water to be added to the ballast tank. Half of the air volume is equal to 42.58m^3 . The weight of the movable part of the pontoon and water in the compression chambers is 26,109t To make the volume and weight correspond to each other it is necessary to find the arithmetic mean of 42,58 +26,109 :2 =34,343. The weight of the pontoon together with ballast water should be 34,343t. So it is necessary to add 34,343t - 26,109t = 8,234t of water. In this case the air volume will be equal to: 42.58m^3 - 8.234m^3 = 34. 346m^3 . The pump will run with the same force for both upward and downward movement P = 34. 343т

As calculated above, the total area of the jets flowing through all nozzles is 301.44cm^2 (6 nozzles of F=80mm each).

The pressure at which the pump will operate with fully open nozzles will be: 4343kg : 301,44cm^2 = 113,929kg/cm^2 or it is 114atm. By reducing the diameter of the outgoing jets, it is easy to get the required pressure in the range of 180 atm -250atm at the same water flow rate.

The capacity of this pumping unit will be calculated according to the formula:
W= flow x head x 9.8m/sec^2 x 0.6 = 1.210m^3 /sec x 220kg/cm^2 9.8m/sec^2 x 0.6 = 1210 litres/sec x 2200m.v.s. x 9.8m/sec^2 x 0.6 = 15652560 W = 15.65MW (at 220atm pressure).

In the case where the working pressure will be equal to 250 kg/cm^2 expected the power will be equal:
W= 1210litres/sec x 2500m.v.s. x 9.8m/sec^2 x 0. 6= 20787000W= 20.787MW

Calculation of the distance from the upper plane of the bulkhead for the water volume in the ballast tank

The base area of the ballast tank is S= 4m x 4m - nr^2 =16m2 (3.14 x 0.31^2 m^2)= 16m^2 -0.937m^2 =15.026m^2 The volume of ballast water required is 8.234t Then the height of the ballast chamber is equal to: (h)

h = 8.234t : 15.026m^2 =8.234m^3 : 15.026m^2 =0.547m

So:- according to the above calculations we have:

The capacity of the two-way pontoon pump is equal to, - 1.21m /sec.^{3}

To work with it in pair, let's choose centrifugal pump D6300-80-1. Its purpose is to supply water to a special tank, in which a two-way pontoon pump works. (see scheme No. 2). This pump has the following operating parameters: Power 2000kW, Head 80m, Supply 6300m^3 /hour or it is equal to6300 :3600 = 1,75 m^3 /sec . With proper adjustment of water supply and its discharge rate into the pontoon pump reservoirs, the performance of the centrifugal pump 1.75m^3 /sec will well cover the performance of the double pass pump which requires 1.21m /sec.^{3}

Sketch #2 shows a schematic of the tank arrangement for operation of a two-way pontoon pump

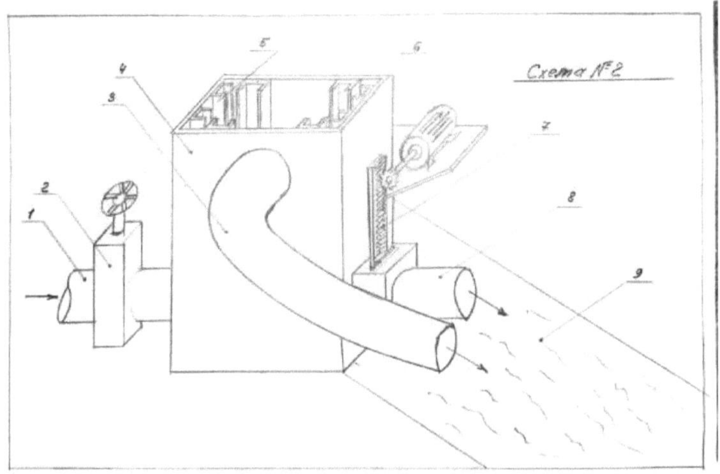

Where: 1) water supply pipeline from the centrifugal pump

2) gate valve for adjusting the water supply
3) overflow pipe above the operating water level
4) tank for operation of the two-way pontoon pump
5) guide channels for roller support guides
6) reversing electric motor-gearbox with adjustable speed from a high-frequency controller.
7) rack-and-pinion opening and closing mechanism to release water and allow water level cycling.
8) water discharge line
9) a concreted channel for general discharge of water into the source water body.

When the proposed design of a pontoon double-way pump with a fixed piston is operating, its maximum power is 20mW. The actual power output will be 18mW (minus the power of the centrifugal pump).

1) For a piston diameter of 299 mm and pontoon dimensions 500mm^500mm4220mm, stroke600mm, the resulting power is 5000W flow rate

water 0.0161m^3 /sec (57.96m^3 /hour) Centrifugal mating pump D160-112 W=15кВт 1500rpm head28m capacity 80m3/hour. In this case there is no useful power output.

2) For piston diameter 500mm and pontoon dimensions 1m-1m-1,8m. stroke 800mm received power 100kW flow rate0,0766m^3 /sec (275m^3 /hour) Centrifugal pump for mating 1D315-71a W=90кВт 3000rpm head60m capacity 300m^3 /hour Useful output power 100kW - 90kW =10kW. When operating the whole installation consisting of two pumps, a total of 10kW of electricity can be obtained.

3) For piston diameter 1196mm and pontoon dimensions 2m^2m^4m, stroke 600mm received power is equal to 3500kW Flow0,111m^3 /sec (399m^3 /hour) Centrifugal pump for mating 1D500-63a W=132kW 1500rpm head 53m productivity 450m^3 /hour Useful output power 3500kW- 132kW= 3368kW =3,4 MW, and it is already essential!

4) According to the above calculations, for piston diameter 2000mm pontoon dimensions according to sketch #1 stroke1,2m Flow1,21m^3 /sec (4356m^3 /hour) Centrifugal pump for mating D6300-80-1 W =2000kW 1000rpm head 80m productivity 6300m^3 /hour Useful output power 20000kW- 2000kW =18000kW =18MW. A plant with such a capacity may well compete with any alternative energy source. From the above data we can draw the following conclusion: that in some cases, when the weight and volume of the pontoon pump is small, it is impossible to obtain excess energy. The greater the weight and volume of a pontoon double stroke pump with a fixed piston, the more energy can be obtained in relation to the energy expended! And the energy expended will be many times and sometimes even orders of magnitude less than the energy obtained!

Similar to the material above, there is another article that explains how to build hydroelectric power plants without blocking rivers.

CHAPTER 3

PROSPECTS OF HYDROPOWER AND LAND RECLAMATION DEVELOPMENT

Azerbaijan Research Institute of Geotechnical Problems of Oil, Gas and Chemistry Author Engineer of Alternative Energy Laboratory
Boris Vladimirovich Silvestrov boris_boris_silvestrov@mail.ru@mail.ru

This article is devoted to the alternative development of hydropower. Traditionally, electricity generation in hydropower is created by building dams, and the capacity is calculated according to the formula

$$W = \text{flow rate x water head x } 9.8 \text{ m}^3 \text{/sec x}0.6.$$

The water flow rate, in turn, depends on the water flow of the river, measured in m^3 /sec. Another factor affecting the amount of electricity generated is the head of the water flow created by the height of the dam. The head is measured in metres of water column. (10m.w.c. =1atm). It is to provide these parameters that a hydraulic structure such as a dam is built. The construction of dams is quite expensive. Besides, this construction, as a rule, is tied to the terrain and not in any point of the river it is possible to erect a dam because of expediency not to flood huge fertile territories. And the number of possible dams on large rivers is very limited. Practically construction of dams on big world rivers has exhausted itself. Flooding of huge fertile lands has become impractical. Dams prevent navigation and migration of valuable fish species. To solve this issue, special hydraulic structures have to be built. Such as locks and bypass channels. This also increases the cost of dam construction and lengthens the payback period of the hydroelectric power plant and does not solve the problems of migration of valuable fish species 100 per cent. The probability of dam failure, although low, still exists, and an accident at the dam can lead to huge man-made consequences. The essence of the proposed solution to the problem is the refusal to build dams. In order to create high pressure, much higher than the dam can create, it is proposed to use the work of the so-called hydropower, gravity module, which is an alternative to the dam in order to create the required head. Whereas conventional hydropower utilises the force of falling water flow, consisting of the head of water and its flow rate, the proposed method uses water in a slightly different way. It carries the function on the basis of which a pontoon piston pump operates. The operation of this pump in turn provides very high pressure and its capacity provides the flow rate of water. The pressure is not created by the falling water, which depends on the height of the dam, but is created by the weight of the moving body of the pump and its air volume, which can activate Archimedes' force. Let us call this device a gravity module. It is a combination of a two-way piston pump with a compensation column and a pressure regulator. This module is able to accumulate energy and release it in the form of a stable stream of water under very high pressure, which will be used to generate electricity or simply to pump water to a sufficiently high altitude, along the banks of the river for irrigation needs. This pump does not require a high water head due to the height of the dam. Electricity is not required for its operation either. Only a small amount of electricity is needed to operate the periodically opening gate to allow the water level to change. It is enough to create a height difference of 2m on the water flow and this is enough for the operation of the gravity module. This difference is created by building a diversion channel or pipeline parallel to the river bed. It can be up to several kilometres. The essence of the work of this module is in the cyclically changing water level. There is a change in the water level in a special tank. This change in the water level leads to the operation of the above-mentioned pump. The pump included in this module has a high capacity, with the ability to create high pressure. This pressure is created also because this pump can be considered as a hydro-press at the same time. A special feature of the design is that the piston is fixed and rigidly fixed in relation to the entire pumping system. But the cylinder is built into the pontoon, which is afloat and has the ability to move relative to the piston in the vertical plane, along with the change in the horizon of the water level. The performance of this pump and the pressure created by it directly depends on the dimensions of the pump and on the height of vertical movement of the cylinder. The cylinder movement itself depends on the water level fluctuation. Compensation column (scheme No.1) is a vessel operating under high pressure. Part of the column is filled with air, which compensates the pulsating character of water supply by the pump to the hydraulic turbine blades.

The pressure regulator (scheme No.1) allows to obtain a relatively stable water jet at the outlet of the compensation column with any given pressure within the maximum possible for a given pump size. Calculations show that for a pump with a piston diameter of 2 m. and

With pontoon dimensions 3.8m x 3.8m x 2m, plus the upper, floating part of the pontoon with dimensions 4m x 4m x 4m, a head of 300 atm is not the limit. The pump is gravity pump because one of the forces working in the vertical plane is the weight of the pontoon with built-in cylinder, and the other applied force is the Archimedes force, determined by the buoyancy of the working pontoon, and is a mirror image of the gravitational force. The geometric dimensions of the pontoon and the material from which it will be made must be selected so that these forces are equal. The operation of the proposed module is related to the cyclic change of the water level horizon. Together with the change of the water level horizon, the working pontoon will also move. The cylinder built into it,

moving relative to the fixed piston, will ensure the operation of the proposed module. The change of the water horizon level can be achieved by using some technical structure. (see diagram No. 1)
This structure is an alternative to a hydraulic structure such as a dam. The purpose of a dam is to raise the level of falling water and thereby store the potential energy to create the required head. And the higher the dam, the greater the head can be generated and the greater the resulting power output of the hydropower plant. In the proposed device, the head is created by the weight and air volume floating the casing of the double pass pump. Where two forces are used are its weight and its buoyancy (Archimedes force). The highest pressure derivative hydroelectric power plant is the one in the Alps, which has an altitude difference of 1883 metres. In other words, this head uses a pressure of just over 180 atm. This is the Bjedron hydroelectric power station in Switzerland. Each of the three hydro turbines, consumes water - 25 m^3 /sec, while the power output of each hydro turbine is equal to 423 MW. The total capacity is 1269 MW. Here is the essence of the proposed method. Such a high head instead of a dam can be created by floating pump casing. And another advantage of the proposed method is that quite powerful hydro power plants can be built on small rivers with water flow from $3m^3$ /sec, and the height difference over a short distance of only 1.5m-2m. For this purpose it is necessary to build an alternative hydro-technical construction to the dam. (see scheme No.1) To achieve this goal it is necessary to construct a water conduit along the river bed, which at the end point will be provided with a height difference of only 1.5-2m. This will allow periodically, by gravity to fill and discharge water in a certain tank, in which the proposed hydropower gravity module will work. Filling and discharging of water can be easily automated by opening and closing the inlet and outlet gates.
The actuation mechanism of the gates depends on reaching the upper and lower working level of the pontoon. The energy required to operate the gates in one case will be stored in the hydropneumatic springs by the movement of the pontoon itself. In the other case the gates will be operated by an electric actuator. To complete the above picture more vividly, let us consider the possibility of hydropower development on the Kura River in the Republic of Azerbaijan. There are several reservoirs on the Kura River, at the outlet of which hydroelectric power plants are built.
It is considered that the Kura, as well as all European and world rivers, has exhausted its resource for construction of large hydroelectric power plants. Loss of fertile lands during construction of water reservoirs and flooding of huge territories, costliness of this construction make such construction inexpedient. But there is a very promising alternative. The average height difference on the

Kura River is 1.8 metres per kilometre. There are many areas where the construction of a conduit or a diversion channel along the river bed along the slope covering the river bed until the height difference of 1.5-2 m is reached would not be very long. Such sections can be found not only upstream of the Shamkir reservoir, but also downstream of Varvara HPP. Downstream of Varvara HPP, the water flow of the Kura River is equal to 402 m^3 /sec. If we build a water conduit or canal along the river bed, with a water flow of 75-80 m^3 /sec, until reaching a height difference of 1.5m-2m, an alternative hydropower plant can be built at the outlet. On the basis of hydroelectric gravity modules, it is possible to obtain the same capacity as the Bjedron HPP, namely 1269 MW, and possibly even higher, as the head can be created much higher than 180 atm (as has already been said, 300 atm is not the limit). And this capacity is almost three times more than the capacity of the Mingichaur HPP, which is equal to 414 MW. At the same time, the above proposed construction does not interfere with navigation and does not harm the ecosystem of the river. And several dozens of such HPPs can be built along the river bed. The cost of such a hydraulic structure is incomparably lower than the cost of traditional dams. In the future, the Mingichaur HPP can be reconstructed and using the water already collected in the reservoir, using a block of hydropower gravity modules to switch to high-pressure bucket hydropower units. Theoretically, a height difference of 40 m allows creating 20 cascades of 2 m each. The total water discharge after Min. HPP 402 m^3 /sec, dividing it by 25 m^3 /sec we get 16 units such as at the HPP in Switzerland at each cascade. Even if we assume that there will be 10 of them on each cascade, the total possible number of hydro turbines involved on the whole cascade will be 200 units each with a capacity of 423 MW. Then the capacity Mingechevir HPP may amount to 84600 MW of electricity. One hydropower module with a capacity of 15MW can be used on the Kura tributaries for every 2m^3 /sec of water flow with a 1.5-2m height difference. The cascade method of construction of such hydropower plants, where water after being used up in one reservoir flows into another one located below, also contributes a huge potential of possible capacities to the total hydropower potential of Azerbaijan. Each involved hydropower module with a piston diameter of 2 metres will produce at least 15 MW of electricity. If we imagine that on the constructed water pipeline Oguz-Gabala-Baku where the total height difference is 227 m and the water flow rate is 5 m^3 /sec and if we imagine that water delivery to the end point will pass through a cascade of hydropower modules each of which has a water flow rate of 1.5 m^3 /sec at the necessary height difference of 2 m, then with three modules installed in parallel at each stage of the cascade, and such stages can theoretically be 113 (227 : 2 =113). Even if we take the number of cascades not

113, and so for a round figure, equal to 50, the total number of modules will be equal to 150 pcs. And thus the total power generation capacity can be 1800 MW of electricity, where each hydropower module with a piston diameter of 2 m has a capacity of 15 MW. Can you imagine that while the main task will be solved, which is the delivery of drinking water to the destination, there will be a huge amount of energy generated in addition! With the use of such alternative hydropower technology, the hydropower potential of Azerbaijan through a set of rivers and canals can fully provide energy not only for the entire republic, for the vast future, but also for the entire Transcaucasus. Since the hydroelectric gravity module is based on the injection of water under high pressure, it can also be used in land reclamation. The same hydraulic structure as in hydropower engineering, but without a compensation column and pressure regulator, will allow pumping water to a considerable height without using any other energy than the potential of the river's height difference. Huge areas adjacent to the riverbed can be irrigated in this way.

Most of the world's countries have signed the Kyoto Agreement, which commits them to reducing their carbon dioxide emissions. But this is just a declaration. The only partial solution to this problem today is alternative energy. Solar, wind, wave and tidal energy are not constant. The factor of cyclicality and dependence on weather conditions do not allow us to fully rely on their development and their further replacement of traditional energy sources. The alternative method of hydropower development proposed above is aimed at solving this problem. The untapped potential of this method is enormous.

Hydropower gravity module in electricity generation is innovative in nature. The demand for such environmentally friendly stable electricity will be very high.

Scheme No. 1 shows the design of the proposed hydraulic structure.

Scheme 1

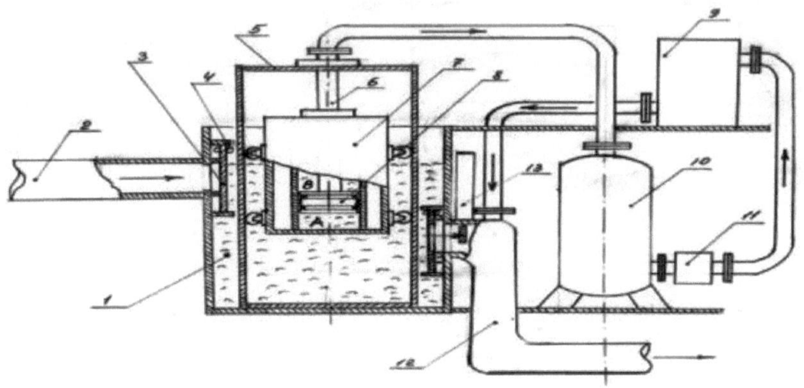

1. Buried concreted tank; 2. Inlet conduit (channel); 3. Inlet shandor (blocking device); 4. Rollers to facilitate the movement of the shandor; 5. Guide grid in which the pontoon pump moves; 6. Conduit; 7. Two-way pontoon pump; 8. Fixed piston; 9. Hydroturbine and generator; 10. Compensation column; 11. Pressure regulator; 12. Waste water discharge conduit; 13.

Shandor opening-closing mechanism on the water discharge.

Water is periodically filled and discharged through the conduit (2) through the operation of the inlet (3) and outlet shandors, changing the water level in the buried tank. The changing water level enables movement of the movable body of the pontoon pump around the stationary piston. This ensures the operation of the pump itself, in which water is "compressed" in chambers "A" and "B" and flows through the water line (6) into the compensation column (10). The regulator (11) supplies water under the set pressure to the hydro turbine (9) where electricity is generated. Waste water is discharged into the water discharge conduit (12)

Fig. No. 2 shows the design of a two-way piston piston unit pontoon pump included in the hydropower gravity module.

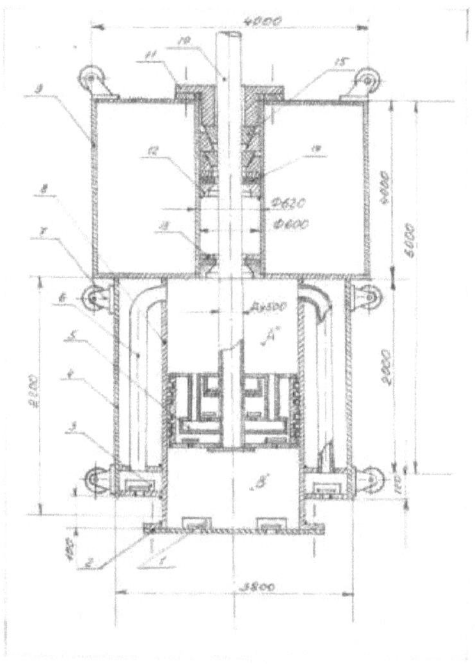

Fig.2

With its design and operation you have familiarised yourself in the previous article, so I will not bore you with repetition of the material.

We have: capacity of the pontoon pump $1.2m^3$ /sec. The water pressure depends on the diameter of the nozzle holes supplying water to the hydroturbine blades and may well reach 250atm. To pump water to the slopes along the riverbed, the required pressure will be such that it exceeds the back pressure created in the conduit.

The capacity of this pumping unit will be calculated according to the formula:

W= flow x head x $9.8m/sec^2$ x 0.6 = 1210kg/sec x $220kg/cm^2$ $9.8m/sec^2$ x 0.6 = 1210 litres/sec x 2200m.h.s. x $9.8m/sec^2$ x 0.6 = 15652560 W = 15.65MW

Let's summarise the above material:

1) We have an alternative method of hydroelectric power generation
2) The potential for utilising hydro resources is increasing manifold.
3) This solves the climate problem
4) It makes it possible to build powerful hydroelectric power plants without building dams on low-water rivers.
5) Allows you to reconstruct the existing hydroelectric power plants in hundreds, thousands of times increasing their capacity.
6) Reduces risks of man-made accidents
7) Provides an opportunity to make a significant breakthrough in the development of the global energy and economy.
8) The proposed method allows, to use the cascade method in construction, G.E.S.
9) Hydropower gets a second wind!

Based on the discovery in the field of physics of a mechanism capable of multiplying energy, a unique source of alternative energy has been designed. This energy source is able to operate in a "perpetual motion" mode. Perpetual motion in terms of energy consumption. Of course, the moving, rubbing parts of this mechanism will sooner or later lead to its failure, but its uniqueness is obvious. You will get acquainted with such a source of energy in the next article:

CHAPTER 4

HARNESSING GRAVITATIONAL ENERGY.

Azerbaijan Research Institute of Geotechnical Problems of Oil, Gas and Chemistry The

author is an engineer of the laboratory on alternative energy

Boris Vladimirovich Silvestrov.

Abstract: This paper discusses a method of converting the gravitational energy into electrical energy. This transformation takes place in a special installation, the device of which is considered in the proposed article. The article considers the physical meaning of the possible increment of energy, in other words, spending little energy, you can get orders of magnitude more. The increment comes from the gravitational field. And also considered the possibility of creating an installation, K.P.D. which is much higher than 100%. Physics has not previously considered the process of possible energy gain, but such a gain is possible in the system of communicating vessels. The article reveals the regularity of energy gain depending on the volume and weight of the working body.

Keywords: Electricity, gravity, communicating vessels, two-pass pump, hydroelectric generator

Modern science denies the possibility of creating a mechanism whose efficiency would be 100% and higher. This denial is based on the fundamental law of physics, namely the law of conservation of energy. Any movement, any work is accompanied by losses on friction, on resistance of the medium. Taking into account these losses, the efficiency will always be less than 100%. Graphically, this process looks like this:

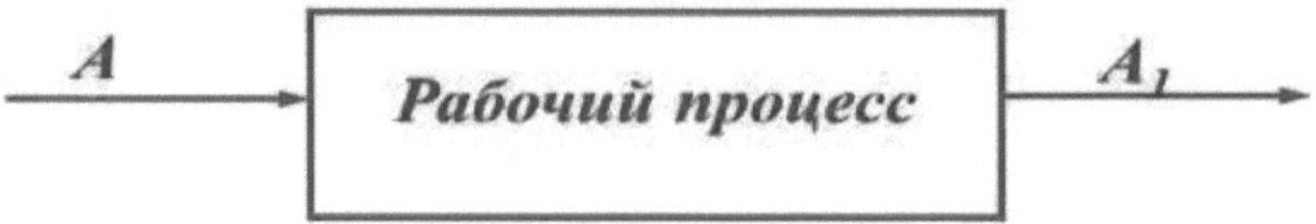

Where, A - energy spent on the working process

A_1 - electricity received

A is always greater than A1 .

But there is a process that has not been considered in physics before, which allows us to conclude that the possibility of obtaining efficiency above 100%, in the conditions of Earth's gravity is quite real, and that this process does not fall under the law of conservation of energy.

This process takes place due to the inflow of energy from the gravitational field. Graphically it can be represented as follows:

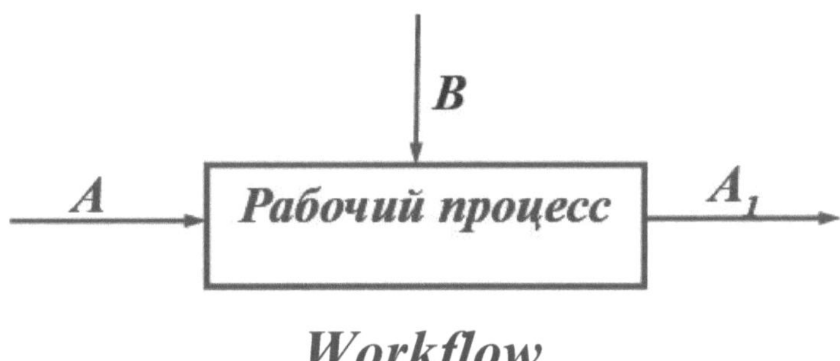

Workflow

Where A - energy spent to change the water level in the communicating vessels; B - gravitational energy, which, in the proposed mechanism is due to the weight and volume of the pontoon pump; A1 - received electric energy.

Thus we have : B > A1 B = A1 B < A1

All three values are valid. It depends on the volume and weight of the floating working body, which comes into play when the water level changes.

There is a process of energy gain due to the fact that to lift a body, in some cases, much less energy is spent than the energy that the body acquires by being lifted to a certain height. It is this phenomenon, this process and will be considered in this article. Gravitational energy has been used by mankind since ancient times. In various mechanisms with counterweights. In clocks - walkers, using the energy of suspended weights, in throwing tools of antiquity, in many other mechanisms, but as a source of electrical energy gravitational field has not been used before. The use of Earth's gravity for energy production is much more promising than wind and solar energy. Since this energy does not depend on external factors, is always stable and can be obtained at any point of the earth, both on the surface and underground, which in turn makes it possible to create "underground cities" both on earth and on other planets. Placed on a railway platform, such an energy source makes it possible to convert all railway transport, and much more, to electric traction!

Let's take a closer look at how this is possible.

Any body of mass m raised to a height h has energy equal to mgh, exactly the same energy (without taking into account losses) will be expended to raise this body to the same height.

But there is another way to move the same body, to a height h, with completely different energy expenditure. For the moment we will not say with greater or lesser expenditure of energy, we will emphasise only that with other! Let us imagine that a body of mass m is afloat. In this state, its weight is zero relative to the water level,

because it is completely compensated for by Archimedes' force. There is absolutely no amount of energy spent on its lifting together with the change of water level. Energy is expended solely to change the level of the water. At the same time many different bodies can be moved alternately, being afloat with different masses, possessing in consequence different energetic possibilities (provided that they will be completely in the air medium) and all this, at the same energetic expenses for change of water level. Hence, in this process the application of the law of conservation of energy becomes meaningless! As the water level changes, the position of the floating body in the vertical plane will also change. In other words, the body can be moved to the height h together with the increase of the water level, where it will acquire energy mgh, provided that it will not be afloat, but will be completely in the air medium. Note that no energy has been directly expended to raise this body of mass m. The energy was expended in a completely different process, namely in the process of changing the level of water, which is provided by immersing and removing a certain weight in one of the communicating vessels. This is the physical meaning of the increment of energy in this process. As in the subsequent process the energy mgh appeared without direct expenditure of energy on lifting, can be taken away in the process of work of the special device providing a condition when the body of mass m completely appears in air medium. In principle, part of this process has been used by mankind long ago in the operation of locks. By changing the water level, we move objects afloat in the vertical plane. Let us set ourselves the task to compare the energy expended in changing the water level and the energy obtained in the accompanying process of lifting a floating body of mass m to a height h. Let us consider this process in the following variant. Let us take a system of communicating vessels consisting of two adjacent tanks, at the base of which lies a square with a side of 1.001m and a height of 2m. The tanks are half filled with water. In the left tank we will alternately immerse and extract a certain body, which is a cube with sides of 1m. The energy for its immersion and extraction will be the energy spent. Theoretically, a small gap allows to perform this operation. In the right tank will be afloat a certain body with dimensions, at the base of which lies a square, with sides of 1m, and the height of this body can be 0,5m., 2m., 3m., 4m. , 3m., 4m., and so on. (see diagram No. 2). The expected work of these bodies is the source of the expected energy production.

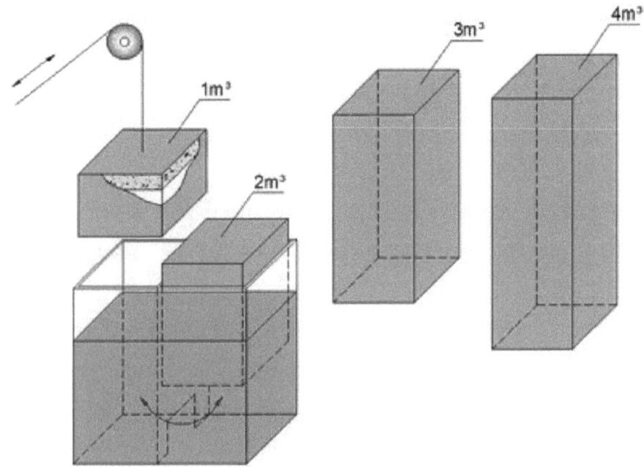

Scheme No. 2

Let us consider the variant with the height of the working body equal to 2m. Let's fill this body with water so that the total weight of water and material from which this body is made would be equal to 1.95t. With the volume of this body equal to $2m^3$ the residual buoyancy will be equal to 50 litres. The body would still float. We will call one immersion and one retrieval of another body changing the water level one cycle. In order to submerge the body changing the water level and having a volume of $1m^3$ must weigh more than 1t let's assume that it weighs 1.001t. The work done by this body in one cycle will be equal to: the arithmetic mean of the weight of this body in the air, and the weight in a fully immersed state, multiplied by the displacement equal to 1m. The arithmetic mean is taken because as this body is immersed, its weight will change, decreasing from 1001kg to 1kg. Conversely, as this body is removed, its weight will increase from 1kg to 1001kg. Although, when travelling downwards, the work can be disregarded because the work is done by the body itself weight, but for the avoidance of doubt, let us take the whole cycle as work. Thus, the work expended for one cycle will be equal to:

A= (1001kg +1kg) / 2 - 1m ^2=1002kgm.

The work done by a body afloat in the right tank, provided that the body is

subsequently entirely in the air, is equal to:
A1= 1950kg -1m -1 = 1950kgm.
The climb phase involves no useful work, so it is multiplied by 1.
Consider the following case, when the height of the body in the right tank is 3m. As in the first case, we fill this body with water, so that the total weight of the material from which it is made and the weight of water is equal to 2950 kg. As in the first case, the residual buoyancy will be equal to 50 litres. ($50dm^3$) Then the work this body could do in one cycle would be equal to:
A2 = 2950kg - 1m - 1 = 2950kgm.
In the same way, consider the case when the height of the body in the right tank is equal to 4m. Let us also fill this body with water until the total weight is equal to 3950 kg. The work that this body could do in one cycle will be equal to:
Az= 3950kg - 1m -1 = 3950kgm.
When considering the variant where the height of the working body is equal to 0.5m, we obtain that the energy expended will be greater than the energy received.

This process has been graphically expressed above:

Where, "B" is the energy expended to change the level of the water table by sinking and retrieving a certain weight. In our example, "B" is const; "A2" is the energy transferred by the working body (pontoon body).
pump). Depends on the volume and weight of the pontoon, and increases as the weight and volume of this working pontoon increases. "A1" is the resulting electricity
From the above calculations, the following rule of thumb can be deduced:
In the system of communicating vessels at the same energy costs for changing the water level by immersion and removal of a certain body in the left tank a significant increase in the energy of the body is possible,
of the floating body in the right tank. This energy gain is directly related to the volume and weight of the floating body in the right tank.
By increasing the volume of the floating body, we get an increase in energy. From this we can conclude that the efficiency increases, which was initially above 100%, provided that the volume of the floating body is greater than the volume of the submerged body. The condition for this process (volume increase) is the increase in the height of the walls of adjacent tanks, the amount of water and the stability of the tanks themselves. But the above example shows only theoretical possibility of energy increase. No matter how much the water level changes and no matter how much the body floating in the right tank changes its position vertically, no increase in energy actually occurs. It is necessary that the considered forces (weights) were activated. And this is possible only by using

some mechanism, which is a two-way pontoon pump with a fixed piston, the work and design of which will be considered below. In view of the above material, we propose to consider a mechanism (installation) converting gravitational energy into electrical energy. (sketch No. 1)

Sketch No. 1

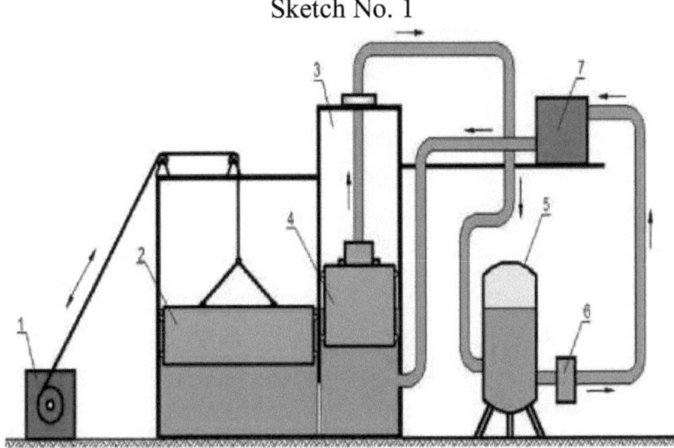

Where: 1 - Reverse electric motor (oil rocker); 2- Weight affecting the change of water level.; 3- Communicating vessels; 4-Two-way piston pump with fixed piston; 5-Compensating column; 6 - Pressure regulator; 7 - Hydrogenerator with Pelton turbine

The operation of this installation is as follows: The reversible electric motor No. 1, or some mechanism like an oil rocker replacing it, alternately immerses and removes a certain weight No. 2 from the water. This makes it possible to change the water level in both tanks. This in turn facilitates the operation of the pontoon pump #4. The pump under pressure from the tank pumps water into the compensation column No.5. (at the moment of start-up of the plant there is a process of adding water to the tanks) Then under the required pressure provided by the pressure regulator No.6, the water goes to the blades of the hydraulic turbine No.7, where electricity is generated. Then water returns to the communicating vessels, after which water addition is stopped. The plant enters the operating mode, is disconnected from the external power source and enters the mode of independent energy supply. The physics of energy increment occurs when the pontoon pump is operating. Therefore, its design and its operation will be discussed in detail. The power of the electric motor for the operation of load No. 2 is selected according to the weight of this load in the catalogue of the hoisting mechanism. The resulting power is calculated from the capacity of the pump and the pressure it can generate according to the formula:

W=Hanop - flow rate - 9.8m/sec^2 - 0.6

The design of the working pontoon-pump represented by bodies **B, C, D on the** scheme No.2 **is** shown on the sketch No.2.

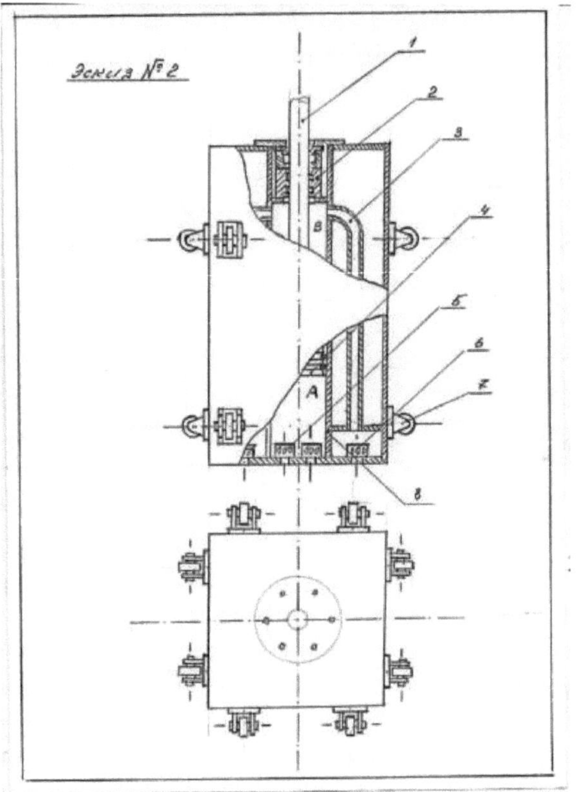

Where :

1 - Fixed water line, at the end of which the piston head (4) is located. The water line is rigidly fixed to the tanks, which are a system of communicating vessels; 2 - Seal assembly; 3 - Pipework for the inlet of the seawater into the upper chamber of the pump **"B"**; 4 - Fixed piston head; 5 - Inlet valves of chamber **"A"**; 6 - Inlet valves of chamber **"B"**; 7 - Support rollers (for ease of movement); 8 - Internal working cylinder.

The operation of this pump is as follows:

As the water level changes, the pump casing moves together with the inner cylinder. For ease of movement, guide rollers are welded to the pontoon body. The cylinder encloses a fixed piston head. During the movement of the pontoon and the cylinder in it, the volumes of chambers "A" and "B" alternately contract and expand. Pressure on the water in one of the chambers is exerted either by the force of weight when moving downwards, or by Archimedes' force when moving upwards. Archimedes' force is a mirror image of the force of weight.

This movement is due to the tendency of the floating pump casing to follow the changing horizon of the water level. Chambers "A" and "B", through a system of inlet valves, receive water from the tank itself, in which the pontoon is located. When the pontoon moves in one of the chambers, water is pressurised by the weight of the pontoon and is forced into the conduit, while the other chamber increases in volume and water is sucked into it. Then the same process takes place in the other chamber. When designing the pump, the weight of the moving part of the pump, by adding water to the ballast chambers located in the upper part of the pontoon body, is selected so that it is equal to the buoyancy (Archimedes' Force) of the pontoon. Since, weight, and buoyancy are equal, the pump operates with the same force, both when travelling upward and downward. The water is returned to the tank (communicating vessel) after being exhausted in the hydraulic turbine. The body of the movable part, the pontoon, can be compared with the above discussed bodies with heights of 0.5m, 2m, 3m, 4m from which the energy increase was planned. But unlike these bodies, when the water level changes, the pontoon is deprived of the possibility of free movement following the changing line of the water level horizon. This movement is prevented by the water in one of the chambers. The pontoon's desire to maintain its equilibrium position in relation to the changing water level line causes the weight force to activate in one case and the buoyancy force in the other. These forces alternately do work. In other words, they give off energy. This energy is expressed as a certain volume of water under a certain pressure.

Sketch **No. 1** shows the valve system of the fixed piston head

WHEREAS:

1) Support area for piston head immobility
2) Water pipe on which the piston head is mounted
3) Piston head
4) Water inlet opening in the conduit
5) Openings for water flow through valves **"A" "B" "C"**

During the upward movement of the working body **"C" (scheme No.2**) the compression chamber "**B" (sketch No.2**) is enlarged and filled with water during the downward movement of the body **"C" the** valves "**B**" and **"C" are** opened and the valve "**A" is** closed. Water through the open valves **"B"** and **"C"** from the compression chamber **"B"** enters the water pipe. At the same time there is an increase in the volume of the chamber **"A"** **(sketch № 2)** and it

is filled with water. Then, when the working body **"C"** moves upwards again (**scheme No.2**), valves "**A"** and **"C"** open and valves "**B"** close (**sketch No.1**). Water, from the working chamber **"A" (sketch №2**), through the open valves

"A" and **"C"**, enters the water line, and further along the technological chain to the blades of the hydraulic turbine.

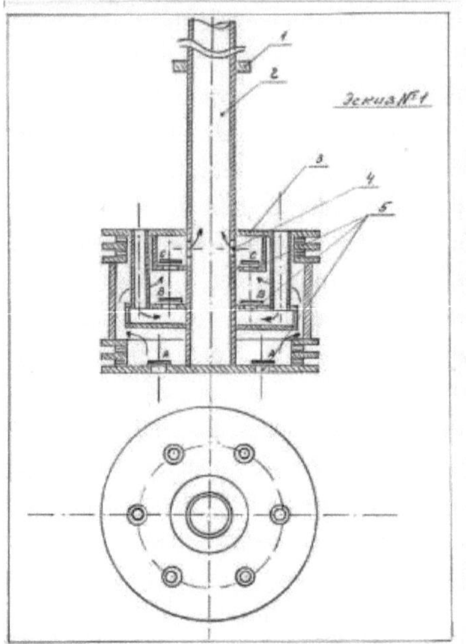

6) Scheme No.1 is nothing else but a gravity energy pump, where the energy increase occurs due to gravity. The larger will be the dimensions of the pontoon-pump, the larger will be the volume of compression chambers of the pump, hence - and productivity of the pump per unit time, and the greater will be the forces (weight force and Archimedes' force) will participate in the production of energy. The greater the applied forces and the greater the capacity, the more energy can be produced at the same energy cost to change the water level horizon.

7) All the theoretical calculations presented are confirmed calculations of the possibility of obtaining electricity, according to the scheme shown in sketch No. 1, and are made according to the formula adopted in hydropower:

W=Hanop - flow rate - 9.8m/sec^2 - 0.6

The head (water pressure) can be found from the ratio of the applied force to the area of the flowing jet. The water flow rate is calculated from the geometric dimensions of the working chambers of the pump and the execution time of one cycle of operation. The applied forces themselves are calculated by determining the weight of all parts and assemblies included in the moving part of the pump and by determining the internal air volume of the moving pontoon, with subsequent equalisation of these forces with ballast water.

8) According to these calculations at different diameters of a piston of a two-way pump and different its dimensions we have the following results: for diameter of a piston 299 mm and dimensions of a pontoon 500mm^500mm4220mm, the spent power (in the form of a lifting device) 540 W, received - 5000 W efficiency=925%.

9) For a piston diameter of 500mm, pontoon dimensions:- 1m-1m-1.8m. 3 kW input power, 100 kW output efficiency=3333%

10) For a piston diameter of 1196mm, the pontoon dimensions are:- 2m^2m^4m, input power is 30 kW received 3500 kW efficiency=11666%

11) For piston diameter 2000mm and pontoon dimensions 3900^3900^8700mm, power input 50kW received 14000kW efficiency=28000%

Physics has not previously considered the process of possible increment of energy. Therefore, the statement that the lifting of a certain body requires less energy than the body will acquire, being lifted to a certain height, is usually rejected with reference to the law of conservation of energy. The above material shows the mechanism of possible increment of energy in the system of communicating vessels, where the application of the law of conservation of energy is meaningless. And the considered process can be considered as a discovery in the field of physics. The process at constant energy costs, which leads to a change in the water level in one of the adjacent tanks of communicating vessels, is a component to the expected energy gain depending on the changing size of the pump. I hope that this article will allow to complete some concepts in the field of physics and to take more seriously the proposed method of transformation of gravitational energy, as well as to recognise in the gravity pump a pseudo perpetual motion machine.

The process occurring in the mechanism, where the expended energy is constant, and the energy obtained as a result of the work of this mechanism varies, depending on the weight and volume of the working body, does not fall under the law of conservation of energy and thus is a mechanism of energy increment. (the words changing and saved already contradict each other) At constant value of the expended work the value of the acquired energy changes, depending on the volume and weight of the lifted, floating body.

With the kind of performance demonstrated by this alternative energy source design, competitiveness with conventional energy sources is quite evident.

Baku 03.09.2020

List of literature used

1) Design and calculation of hydraulic turbines [Text]. - 2nd ed., supplement. and revision. - Leningrad : Mashinostroenie. Leningr. department, 1974. - 408 c. : ill.; 26 cm.
On the back of the title page authors: S. A. Granovsky, V. Malyshev. M. M., Orgo V. M., Smolyarov L. G.
RuMoRGB
1st ed.: S. A. Granovsky, V. M. Orgo, L. G. Smolyarov. Hydraulic Turbine Designs and Calculation of Their Parts RuMoRGB
Hydraulic turbines - Design and calculation
Storage Cipher:
FB B 74-16/108
FB B 74-16/109
FB Arch.
2) KYRGYZ-RUSSIAN SLAVIAN UNIVERSITY Department of Hydraulic Engineering and Water Resources N.P. Lavrov, G.I.
Loginov Designing of hydro-equipment of DERIVATION hydroelectric power station and selection of the basic power equipment Methodical instructions for course and diploma design.
3) Ministry of Education and Science of the Russian Federation Federal State Budgetary Educational Institution of Higher Professional Education "Nizhny Novgorod State University of Architecture and Civil Engineering" Fevrelev Arkady Valentinovich PROJECTING HYDROELECTRIC STATIONS ON SMALL RIVERS Approved by the Editorial and Publishing Board of the University as a textbook.
4) C.Mizner, C.Thorne and J.Wheeler, Gravity.(can be used mainly - in places - as a reference book, e.g. Chapters 31-34). 31-34).
5) Vlasov V.N. Complexity and simplicity of our existence - 14. Secrets of gravitation and magnetism - at the service of the Russian World!
6)Chernyaev I. A. Piston pumps. M.: Mashinostroenie, 1966. - 188 c.
7) Internet site
Pelton hydroelectric turbine generator....
fstgenerator.com>en/hydroelectric-pelton-product/

yes
I want morebooks!

Buy your books fast and straightforward online - at one of world's fastest growing online book stores! Environmentally sound due to Print-on-Demand technologies.

Buy your books online at

www.morebooks.shop

Kaufen Sie Ihre Bücher schnell und unkompliziert online – auf einer der am schnellsten wachsenden Buchhandelsplattformen weltweit! Dank Print-On-Demand umwelt- und ressourcenschonend produzi ert.

Bücher schneller online kaufen

www.morebooks.shop

info@omniscriptum.com
www.omniscriptum.com

Printed by Books on Demand GmbH, Norderstedt / Germany